Micromanufacturing Operations

Selim Gürgen
Editor

Micromanufacturing Operations

Springer

Editor
Selim Gürgen
Department of Aeronautical Engineering
Eskişehir Osmangazi University
Eskişehir, Türkiye

ISBN 978-3-032-04192-0 ISBN 978-3-032-04193-7 (eBook)
https://doi.org/10.1007/978-3-032-04193-7

This Springer imprint is published by the registered company Springer Nature Switzerland AG
The registered company address is: Gewerbestrasse 11, 6330 Cham, Switzerland

Contents

Chapter 1
Micromachining Processes for Additively Manufactured Metallic Parts

Mehmet Alper Sofuoğlu

Abstract This chapter discusses the micromachining processes applied to metallic parts produced through additive manufacturing (AM) technologies. While AM offers significant advantages in producing complex geometries and lightweight components, it often results in challenges related to surface roughness and dimensional accuracy. As such, micromachining serves as a critical postprocessing step to achieve the desired precision and surface quality of AM parts. Techniques such as micromilling, microdrilling, and micro-electrical discharge machining (micro-EDM) are frequently employed for processing AM-produced parts. However, the unique microstructural and mechanical properties of AM metals, as compared to conventionally manufactured parts, introduce various challenges in micromachining. This chapter provides an overview on micromachining methods applied to AM metallic components, addressing issues like surface integrity, and dimensional precision.

Keywords Additive manufacturing · Micromachining · Surface quality · Dimensional accuracy · Hybrid manufacturing

1.1 Introduction

Additive manufacturing (AM) is a recent technology that enables the layer-by-layer construction of parts directly from 3D digital models, offering unparalleled design flexibility and the ability to produce complex geometries that are difficult or impossible with traditional manufacturing methods. One of the key advantages of AM is the reduction in material waste, as material is added only where needed, eliminating the need for extensive subtractive processes and minimizing the generation of scrap. This characteristic also makes AM a more environmentally friendly option, with the potential for significant energy savings and lower carbon footprints compared to conventional manufacturing techniques. Furthermore, AM allows for rapid prototyping and the production of customized parts, making it ideal for sectors such as

M. A. Sofuoğlu (✉)
Department of Mechanical Engineering, Eskişehir Osmangazi University, Eskişehir, Türkiye
e-mail: asofuoglu@ogu.edu.tr

S. Gürgen (ed.), *Micromanufacturing Operations*,
https://doi.org/10.1007/978-3-032-04193-7_1

aerospace, medical, and automotive industries, where precision and tailored solutions are critical. The ability to fabricate lightweight components with internal structures, such as lattice geometries, further expands the scope of applications for AM, particularly in high-performance engineering [1–5].

AM processes, while offering significant advantages, are not without challenges, particularly in terms of the quality of the produced parts. One of the primary issues is the surface roughness, which is generally higher in AM parts compared to conventionally manufactured components, largely due to the layer-by-layer construction process. Dimensional inaccuracies also frequently occur, as AM can struggle to maintain tight tolerances, leading to geometric defects such as warping, stair-stepping effects, and shrinkage cavities. Another significant challenge is the microstructural anisotropy inherent in AM parts, as the directional nature of layer deposition results in nonuniform mechanical properties across different orientations. This anisotropy can adversely affect the performance and reliability of parts, particularly in applications requiring uniform strength and durability. Overcoming these limitations typically requires additional postprocessing, such as heat treatment or machining, to improve surface quality, dimensional accuracy, and microstructural consistency [6–13].

Micromachining is a crucial postprocessing technique for parts produced by AM, primarily to address the limitations in surface finish and dimensional accuracy inherent in AM processes. Due to the layer-by-layer deposition characteristic of AM, parts often exhibit poor surface quality, with higher roughness and geometric inaccuracies compared to traditionally manufactured components. Micromachining is essential for refining these surfaces to meet the stringent tolerances and smoothness required in many applications. Furthermore, AM parts often suffer from anisotropic microstructures and dimensional deviations, which can negatively affect their performance. Micromachining helps in achieving the desired precision and enhancing the overall functionality of these parts by removing excess material and improving dimensional conformity. As a result, micromachining is not only necessary for improving the aesthetic and mechanical properties of AM components but also for making them suitable for critical applications in industries such as aerospace, medical, and automotive [14, 15].

The objective of this chapter is to provide a comprehensive understanding of micromachining as an essential postprocessing technique for improving the quality of metallic components produced by AM. The chapter begins with an overview of the main AM technologies used for fabricating metallic materials, setting the foundation for understanding the unique challenges associated with these processes. It then emphasizes the importance of micromachining as an essential postprocessing technique in addressing issues related to surface finish and dimensional accuracy, which are critical for the functional performance of AM parts. Various micromachining techniques, such as micromilling, microdrilling, and micro-electrical discharge machining (micro-EDM), are introduced and explained in detail. The chapter also discusses process mechanisms and outcomes reported in the literature to illustrate the effectiveness of micromachining in enhancing the performance and reliability of AM metallic parts.

1.2 Additive Manufacturing Technologies

Selective laser melting (SLM) is one of the most widely used AM techniques for metals, particularly known for its ability to produce highly complex, high-performance components. In the SLM process, a high-energy laser is used to selectively melt and fuse metallic powder layers based on a digital 3D model, creating fully dense parts with fine details. As shown in Fig. 1.1, the laser selectively scans the surface of the powder bed, melting and solidifying the metal powder layer by layer until the entire part is formed. This technique is highly suitable for producing intricate geometries and features such as internal channels, lattice structures, and thin walls, which would be difficult to achieve with traditional manufacturing methods [16, 17]. SLM is particularly effective for processing high-strength metals like titanium alloys (e.g., Ti–6Al–4V), nickel-based superalloys (e.g., Inconel 718), and stainless steels, which are commonly used in industries such as aerospace, automotive, and medical devices. One of the key advantages of SLM is the ability to produce near-net-shape parts with minimal material waste and without the need for additional tooling. However, the process has some limitations, as parts produced by SLM often exhibit issues such as residual stresses, anisotropy, and surface roughness, necessitating postprocessing techniques like micromachining to achieve the desired surface quality and dimensional accuracy. Despite these challenges, SLM continues to be a leading technology in metal additive manufacturing due to its versatility and ability to produce high-performance, customized components [18].

Electron beam melting (EBM) is another prominent AM technique for metals, particularly suited for high-performance applications requiring robust material properties and intricate geometries. In EBM, a focused electron beam is used as the energy source to selectively melt metallic powder layers in a vacuum environment, based on a digital 3D model. Figure 1.2 shows an illustration of EBM process designed for producing AM parts. Similar to SLM, the process builds parts layer by layer, but an electron beam allows for deeper penetration and more efficient energy transfer, making it highly effective for processing high-melting-point materials such as titanium alloys (e.g., Ti–6Al–4V) and cobalt-chromium alloys [20, 21]. One of the key

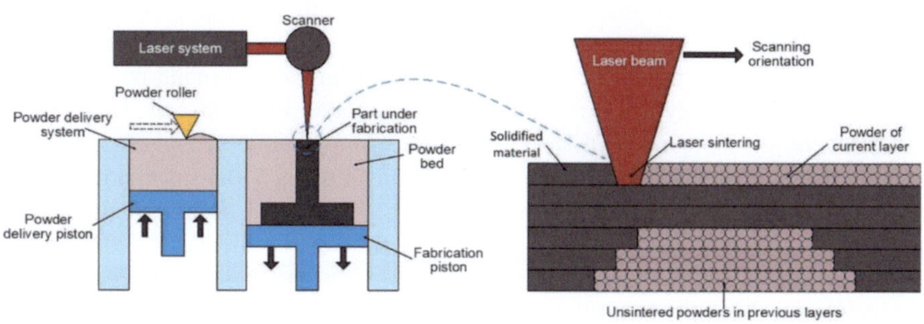

Fig. 1.1 Selective laser melting process [19]

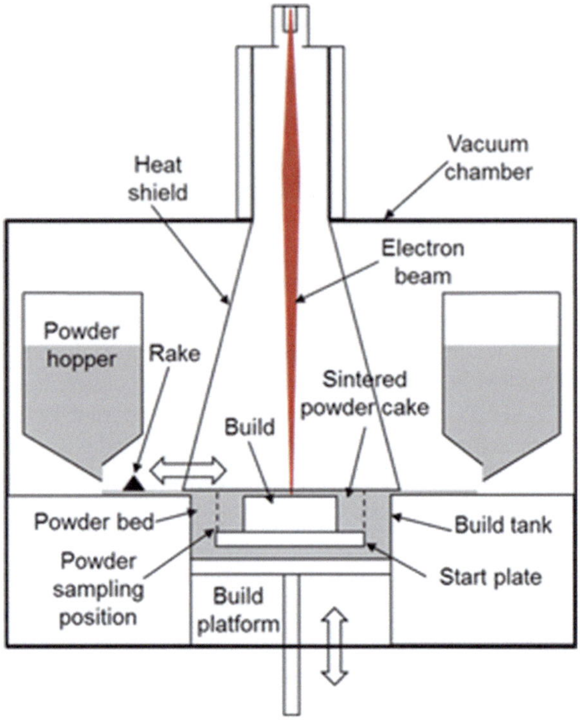

Fig. 1.2 Electron beam melting process [24]

advantages of EBM is its ability to produce parts with excellent mechanical properties and reduced residual stress due to the high temperatures involved in the process, which allow for better control of thermal gradients and grain structure during solidification. EBM is commonly used in the aerospace and medical industries, especially for producing lightweight, complex structures such as turbine blades and orthopedic implants. However, the vacuum environment required for the electron beam process limits the range of materials that can be used, and the surface roughness of EBM-produced parts is often higher than desired, necessitating postprocessing techniques such as micromachining to achieve smooth surfaces and dimensional accuracy. Despite these limitations, EBM offers significant advantages in terms of build speed and material efficiency, particularly for large-scale metal parts, making it a valuable technology in the metal additive manufacturing landscape [22, 23].

Direct energy deposition (DED) is a versatile AM technique that involves simultaneously depositing and fusing material layer by layer using a focused energy source, typically a laser, electron beam, or plasma arc. Figure 1.3 shows an illustration of DED process. Unlike powder bed fusion methods (like SLM and EBM), DED feeds metal in the form of powder or wire directly into the melt pool created by the energy source. This process allows for the precise control of material deposition, making it particularly useful for applications such as repair, part enhancement, and producing large, complex structures [25]. DED has an ability to build or repair parts

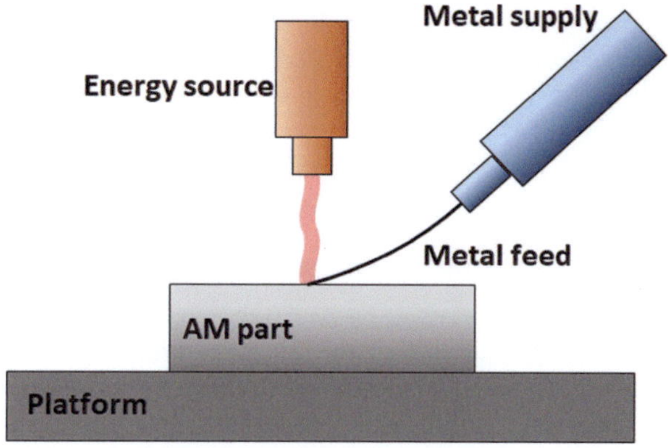

Fig. 1.3 Direct energy deposition process

in a near-net-shape form, which reduces material waste and allows for the production of components with complex geometries and internal features. It can be used with a wide variety of metals, including titanium, stainless steel, and nickel-based superalloys. The ability to deposit material in specific areas makes it an ideal solution for adding features to existing parts or repairing worn or damaged components, especially in industries such as aerospace and defense. However, DED faces challenges related to surface roughness and dimensional accuracy, similar to other AM techniques. The relatively rough surfaces and potential porosity in DED parts often require significant postprocessing, including machining and heat treatments, to meet the high standards required in critical applications. Despite these challenges, DED's flexibility, scalability, and ability to process a broad range of materials make it a valuable technology for both the fabrication and repair of metal components [26, 27].

AM processes can accommodate a wide range of materials, with metals being particularly critical for high-performance applications in industries such as aerospace, automotive, medical, and defense. Among the most commonly used metals in AM are titanium alloys, nickel-based superalloys, and stainless steels, each offering unique advantages for different applications. Titanium alloys, particularly Ti–6Al–4V, are widely used in AM due to their high strength-to-weight ratio, corrosion resistance, and biocompatibility, making them ideal for both aerospace components and medical implants. Ti–6Al–4V is frequently processed using SLM and EBM, producing parts with excellent mechanical properties, although challenges such as residual stress and anisotropy must be addressed through postprocessing. Nickel-based superalloys, such as Inconel 718 and Inconel 625, are also popular in AM due to their ability to withstand extreme temperatures and corrosive environments. These materials are critical in high-temperature applications, particularly in the aerospace industry for turbine blades and engine components. Nickel alloys are often processed using SLM and DED techniques, offering excellent material performance, though their high hardness can lead to difficulties in postprocessing, especially micromachining. Stainless steels, such as AISI 316L, are another common choice for AM, offering good mechanical properties, corrosion resistance, and ease of processing. These materials are frequently used in the medical, tooling, and automotive sectors. Stainless steel components produced via AM, particularly using DED and SLM, often require postprocessing to improve surface finish and dimensional accuracy [14, 15].

1.3 Micromachining as a Postprocessing Operation

Micromachining refers to the precision removal of material on a microscale, typically using mechanical, electrical, or laser-based tools to create fine features and achieve high surface quality in components. In the context of AM, micromachining is an essential postprocessing technique that enhances the surface finish and dimensional accuracy of metal parts, which are often left with rough surfaces and imperfections due to the nature of the layer-by-layer build process. One of the key reasons why micromachining is critical for AM parts is the inherent limitations of AM techniques in achieving the final desired part quality. AM processes such as SLM and EBM can produce complex geometries, but the surface roughness and geometric defects, such as stair-stepping and porosity, often require additional finishing. Micromachining allows manufacturers to refine these surfaces, reducing roughness and eliminating imperfections to meet the stringent requirements of industries like aerospace, medical devices, and automotive. Furthermore, AM parts frequently exhibit dimensional inaccuracies due to factors like thermal distortions and shrinkage during the build process. Micromachining enables precise control over the final dimensions of the part, ensuring that it meets tight tolerances and performs reliably in critical applications. The ability to create microscale features and enhance the overall quality of AM-produced components makes micromachining an indispensable step in the production of high-performance parts across various industries [28–30].

1.3.1 Micromachining Techniques for AM Metallic Parts

1.3.1.1 Micromilling

Micromilling is one of the most widely preferred micromachining techniques, particularly for postprocessing metal parts produced by AM. It involves the use of mechanical cutting tools with diameters typically less than 1 mm to remove material and create fine, precise features on the surface of a part. The process is similar to conventional milling, but it operates at a much smaller scale, requiring advanced control over the cutting forces, tool paths, and machining conditions to achieve the desired level of accuracy and surface finish [31]. In the context of AM, micromilling is essential for improving the surface quality of parts that may have rough or uneven surfaces due to the layer-by-layer construction inherent in AM techniques like SLM or EBM. The fine cutting tools used in micromilling can access small, intricate areas

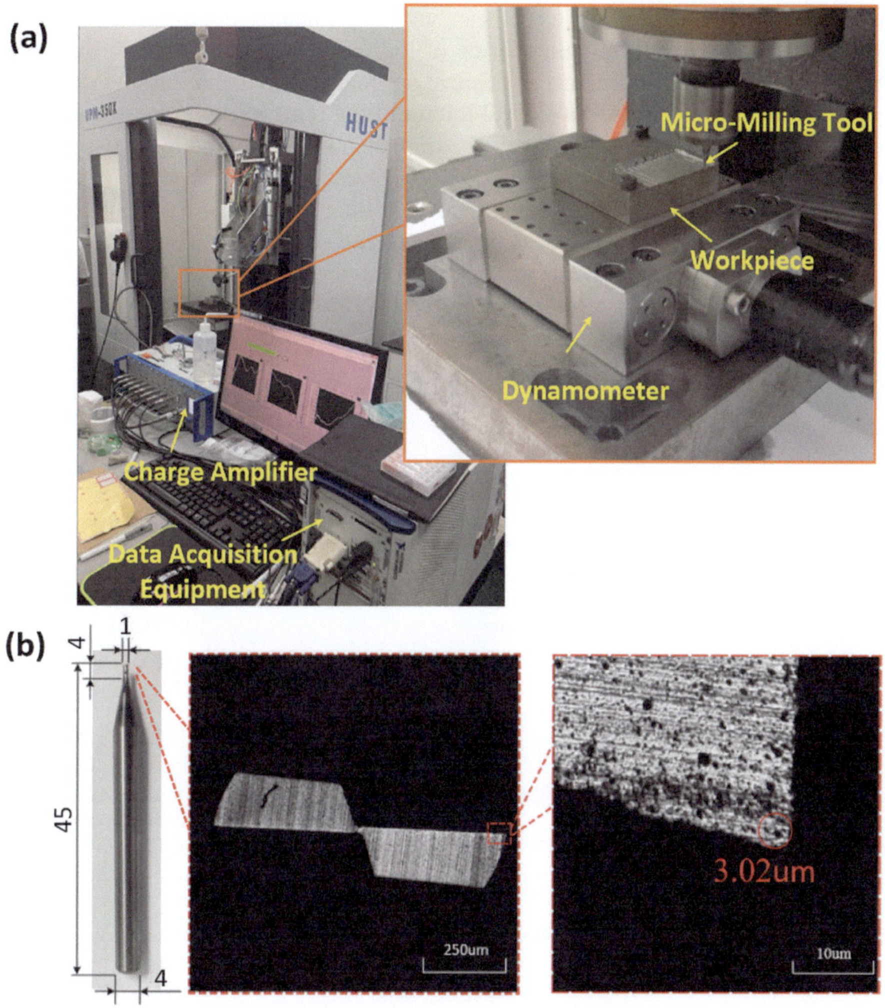

Fig. 1.4 **a** Micromilling setup and **b** micromilling tool dimensions [31]

of a part that larger tools cannot, making it ideal for finishing complex geometries or detailed features that are typical of AM components. Micromilling is particularly useful for enhancing the dimensional accuracy of AM parts. AM processes often leave parts with geometric inaccuracies such as warping or slight misalignments. Through micromilling, these imperfections can be corrected, bringing the part into compliance with tight tolerance requirements [32]. Figure 1.4 shows a micromilling setup and micromilling tool dimensions.

Micromilling of AM parts presents its own challenges. Due to the unique microstructure and material properties of AM metals, such as anisotropy and residual stress, tool wear and cutting forces can vary significantly compared to conventional materials. Therefore, optimization of tool geometry, cutting conditions, and machining strategies is necessary to achieve optimal results in micromilling for AM applications. Despite these challenges, micromilling remains a vital postprocessing method for refining AM parts to meet the high standards required in industries like aerospace, medical devices, and precision engineering.

Table 1.1 summarizes the studies on micromilling of metallic parts produced by AM, highlighting how AM process parameters and micromilling conditions affect machining performance. Materials such as Ti–6Al–4V, stainless steels (SS316, AISI 316L, 17-4 PH SS), Co–Cr–Mo alloy, NiTi alloy, AA4043, Inconel alloys (IN718,

Table 1.1 Micromilling studies

References	AM parameters	Micromilling parameters	Materials	Results
[33]	SLM: no specific AM parameters provided	Tool: PCD micro mill ($\phi = 4000$ µm, 2-flute), $N =$ 10–40 m/min, $d =$ 5–20 µm, $fz =$ 1–4 µm	Ti–6Al–4V	Cutting forces peaked at 20.4 N over 11,400 mm of cutting length, with a tool wear area reaching 10,120 µm^2
[34]	PBF: $P = 200$ W	Tool: carbide ball nose micromills (ϕ = 600, 800 µm), N = 32,000 rpm, $d =$ 100 µm, $fz =$ 0.5–3 µm	SS 316	The 800 µm tool yielded a surface roughness (Ra) between 0 and 0.18 µm. A 120° toolpath produced better geometry, and down-milling improved surface smoothness
[35]	No specific AM parameters provided	Tool: AlCrN coated WC ($\phi = 1$ mm, 2-flute), $Vc = 17.7$, 35 m/min, $fz =$ 0.015, 0.03 mm/tooth, $d = 0.25$, 0.5 mm	AISI 316L	Increasing the feed rate and depth of cut caused a 40% rise in cutting force. Heated parts had a smoother Ra of 1.5 µm compared to 2.7 µm in cold parts
[36]	SLM: $P = 50$ W, V = 300 mm/s, $h =$ 50 µm, $t = 30$ µm	Tool: TiN coated WC ($\phi = 800$ µm, 2-flute), $Vc = 40$ m/min, $d = 200$ µm, fz = 0.5–10 µm	17–4 PH SS	Specific cutting forces in the ploughing regime were 10 times higher than in the shearing regime. Low feed rates increased surface roughness and burr size

(continued)

Table 1.1 (continued)

References	AM parameters	Micromilling parameters	Materials	Results
[37]	SLM: $P = 45$ W, $V = 160$ mm/s, $t = 30$ μm	Tool: SECO 905,004-MEGA-T ($\phi = 400$ μm, 2-flute), $Vc = 27.3$ m/min, $d = 10$ μm, $fz = 0.5$–10 μm/tooth/rev	Co–Cr–Mo	Minimum uncut chip thickness (MUCT) was achieved at a feed rate of 2–3 μm/tooth/rev. Hardness and chip length increased as the feed rate was raised
[38]	SLM: $P = 120$ W, $V = 800$ mm/s, $t = 40$ mm, $h = 100$ mm	Tool: carbide tool ($\phi = 2$ and 0.2 mm, 2-flute), $N = 60{,}000$ rpm, $d = 10$ μm, $fz = 300$ mm/min	NiTi	Machining errors were minimal (0.7 and 1.51%), with cutting forces staying below 1 N. No significant tool wear observed
[39]	WAAM: voltage = 10.7 V, $I = 42$ A, head speed = 5 mm/s, wire feed rate = 2.54 m/min, no. of layers = 40, ILCT = 5,120 s	Tool: WC ($\phi = 500$ μm, 2-flute), $Vc = 80$ m/min, $fz = 1$ mm/tooth, $d = 100$ μm, $L = 5$ mm	AA4043	The sample subjected to a 120-s ILCT experienced lower cutting force, surface roughness (Ra), tool wear, and burr formation
[40]	SLM: $P = 210$ W, $V = 750$ mm/min, $h = 0.1$ mm	Tool: NAlTi-cemented carbide ($\phi = 0.3$ mm, 2-flute), $N = 25{,}000$ and 30,000 rpm, $fz = 5$, 10 mm/min, $d = 0.2$ mm	IN718	SLM IN718 samples exhibited reduced residual compressive stress and 40–45% lower tool wear compared to wrought IN718 samples
[41]	SLM: $P = 160$ W, $h = 110$ μm	Tool: AlTiN-WC ($\phi = 508$ μm, 2-flute)	IN718 (15 × 20 × 5)	AlTiN/Si_3N_4 coating reduced tool wear and improved surface roughness (Ra) to 1–2 μm. Tool life was more affected by cutting speed than feed rate
[42]	SLM: $P = 370$ W, $V = 1200$ mm/s, $S = 170$ μm, $t = 60$ μm, $h = 110$ μm	Tool: TiNi-coated WC ($\phi = 800$ μm, 2-flute), $Vc = 40$ m/min, $d = 200$ μm, $fz = 0.5$–10 μm/rev	IN625	Accurate prediction of cutting force components, with error rates ranging from 0.8 to 14.6%
[43]	Extrusion-AM: nozzle diameter = 0.8 mm, $Vc = 40$ mm/s, $t = 0.6$ mm	Tool: carbide ($\phi = 0.1$ mm), $N = 24{,}000$ rev/min, $d = 5$ μm, $fz = 0.1$ mm/min	Zirconia toughened alumina	Achieved good geometric accuracy and surface integrity, comparable to diamond grinding

IN625), and zirconia toughened alumina were investigated. The studies demonstrate that factors like cutting speed, feed rate, depth of cut, and tool geometry significantly influence outcomes such as cutting forces, tool wear, surface roughness, and machining accuracy. For example, higher feed rates and depths of cut generally increase cutting forces and tool wear, while optimized parameters can improve surface finish and reduce burr formation. Moreover, the studies reveal that specific AM parameters can affect material properties like residual stress and microstructure, which in turn impact micromilling performance. Overall, these findings suggest that careful selection of both AM and micromilling parameters is crucial for enhancing the manufacturability and quality of AM components [33–43].

1.3.1.2 Microdrilling

Microdrilling is a micromachining technique used to create small, precise holes in materials, typically with diameters ranging from a few microns to a few millimeters [44]. Microdrilling serves as a critical postprocessing method, enabling the creation of fine features such as holes, channels, and internal passageways that cannot be accurately produced during the AM build process. This process is particularly important for components requiring high precision, such as those used in medical devices, aerospace applications, and microfluidic systems [45]. An experimental setup is given in Fig. 1.5.

Microdrilling is highly effective for postprocessing AM parts, because AM techniques often produce parts with poor surface quality and dimensional inaccuracies, especially in intricate areas such as small holes or complex internal features. The inherent limitations of AM, including thermal distortions and rough surfaces, make precise hole formation difficult, and microdrilling is required to achieve the necessary precision and surface finish. The process uses ultrasmall, sharp drill bits, typically made from high-performance materials such as tungsten carbide or diamond-coated tools, to remove material while maintaining a high level of accuracy. Microdrilling allows for the creation of deep, small-diameter holes with minimal deviation and high aspect ratios, which is critical in applications where space constraints and precision are paramount. However, microdrilling on AM-produced metals can present some challenges, including tool wear, burr formation, and difficulty in maintaining accuracy due to the anisotropic and sometimes porous microstructure of AM parts. As a result, careful selection of tool material, optimization of cutting parameters, and the use of cooling or lubrication techniques are often necessary to achieve the best results. Despite these challenges, microdrilling remains a good tool for refining and finishing AM parts to meet stringent technical requirements [45].

Microdrilling studies investigate the microdrilling of materials produced by various AM techniques, emphasizing the impact of AM parameters and drilling conditions on machining performance and hole quality. In a study by Sorgato et al. [46], Ti–6Al–4V manufactured via Laser Powder Bed Fusion (LPBF) was drilled using uncoated tungsten carbide drill bits. Higher cutting speeds and feed rates resulted in minimal cylindricity errors, and vibration-assisted drilling enhanced hole

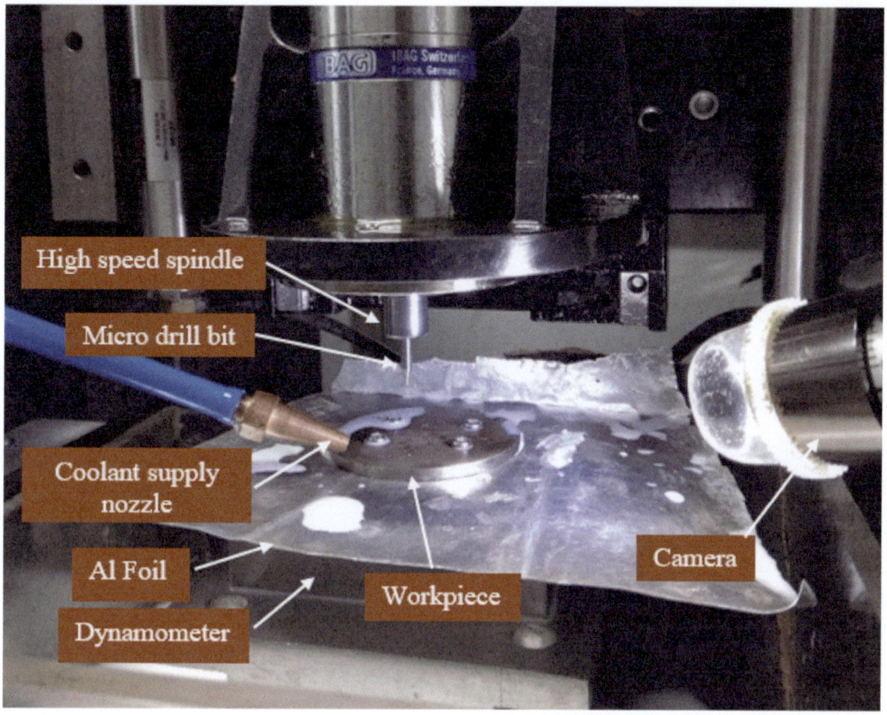

Fig. 1.5 Experimental setup for micro drilling of Incoloy 825 [44]

quality compared to conventional drilling. Rysava et al. [47] also focused on Ti–6Al–4V, revealing that drilling precision was optimal at a cutting speed of 60 m/min and a feed rate of 10 μm, although larger exit burrs were observed compared to entry burrs. In a recent study [48], microdrilling of maraging steel produced by SLM demonstrated that cryogenic treatment effectively reduced material anisotropy, and drilling parallel to the build direction required less energy. Collectively, these studies highlight the significance of optimizing drilling parameters and considering material properties influenced by AM processes to improve microdrilling performance and hole quality.

Joshy and Kuriachen [49] investigated the machinability of AlSi10Mg components fabricated by Laser Powder Bed Fusion (LPBF), focusing on the effects of cutting parameters and build orientation during microdrilling operations. The study revealed that higher feed rates led to increased cutting forces; however, surface roughness remained largely unaffected regardless of whether drilling was performed parallel or perpendicular to the build direction. Nevertheless, build orientation was identified as a significant factor influencing hole diameter and roundness. Overcut was observed in both drilling orientations, attributed to chip adhesion and interrupted chip flow during the process. Furthermore, delamination defects were noted in holes drilled perpendicular to the build direction, as shown in Fig. 1.6. This delamination,

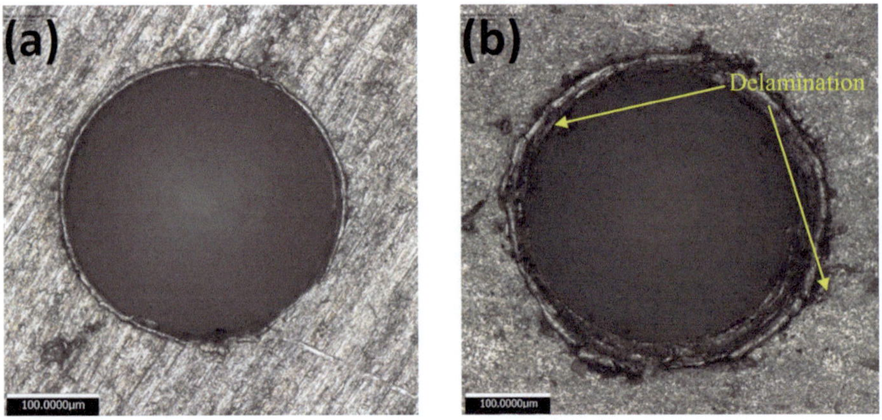

Fig. 1.6 Hole conditions after microdrilling for **a** parallel to the build plane direction and **b** perpendicular to the build direction [49]

referred to as peel-up delamination, occurred at the hole entry and was associated with elevated radial forces acting on the upper layers of the additively manufactured parts.

1.3.1.3 Micro-electrical Discharge Machining (Micro-EDM)

Micro-EDM is a nontraditional micromachining process that utilizes electrical discharges to remove material from a workpiece [50]. Unlike mechanical micromachining techniques such as micromilling and microdrilling, micro-EDM does not rely on physical cutting forces but instead erodes material through a series of controlled electrical sparks between a tool and the workpiece submerged in a dielectric fluid. This makes micro-EDM particularly well-suited for machining hard and difficult-to-machine materials, such as those commonly used in AM parts, including titanium alloys, nickel-based superalloys, and hardened steels. In the context of AM, micro-EDM is a highly effective postprocessing method for improving surface quality and creating precise, intricate features such as microholes, cavities, and complex internal geometries that may be difficult to achieve through conventional machining. Given the layer-by-layer construction of AM parts, which often results in surface roughness, thermal distortions, and geometric inaccuracies, micro-EDM is advantageous for refining these parts to meet stringent dimensional tolerances and achieve high surface integrity. One of the significant benefits of micro-EDM is its ability to machine extremely small features with high precision, regardless of the material hardness. This makes it ideal for postprocessing AM parts where conventional mechanical processes might struggle due to the material properties or complexity of the design. Moreover, since there are no direct contact forces between the tool and the workpiece, micro-EDM minimizes tool wear and is less likely to introduce

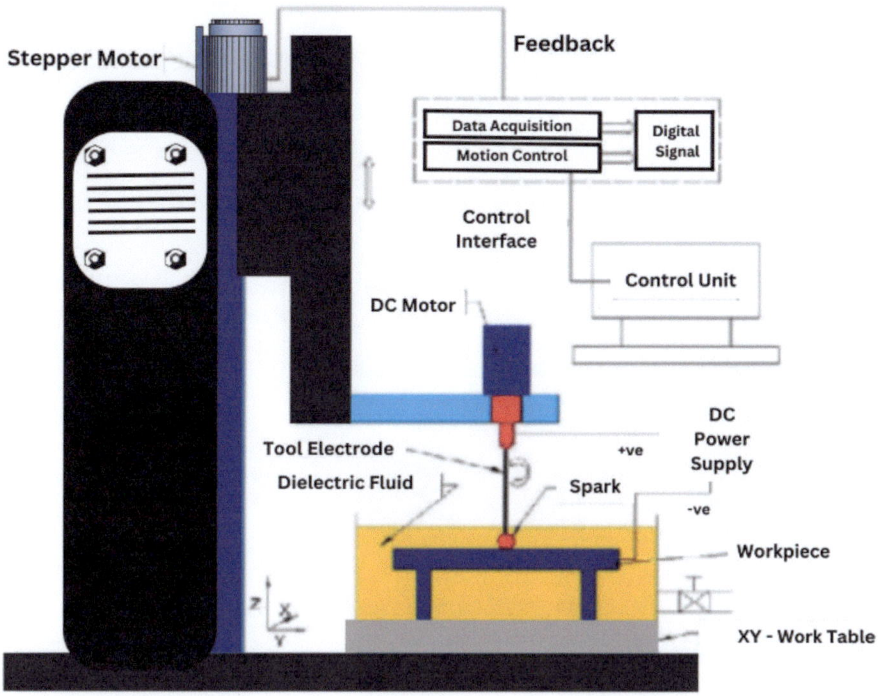

Fig. 1.7 Micro-EDM process [50]

new mechanical stresses or deformations to the part, making it especially useful for delicate AM components. Figure 1.7 shows a schematic illustration of micro-EDM process.

Despite its numerous advantages, the micro-EDM process is inherently slower than conventional mechanical micromachining techniques, which limits its material removal rate. Furthermore, optimal results in micro-EDM require precise control of process parameters, such as discharge energy, pulse-on duration, and the selection of suitable dielectric fluids to minimize thermal damage to the workpiece. Nonetheless, micro-EDM remains a vital technique for enhancing the precision and surface integrity of AM parts, particularly in high-value sectors such as aerospace, medical devices, and microelectronics, where dimensional accuracy and surface quality are of critical importance [51]. In a recent study by Ozaner et al. [52], Inconel 939 fabricated via Laser Powder Bed Fusion (LPBF) was machined using a copper wire electrode. The investigation highlighted that pulse-on duration was the most influential factor affecting machining performance. Similarly, Gokcekaya et al. [53] explored the effects of micro-EDM on LPBF-manufactured Inconel 718 using a brass-coated copper wire electrode. Three cutting strategies, namely, rough, semifinish, and finish, were employed to investigate the surface quality outcomes. Rough machining involved a single pass, while semifinish and finish strategies used

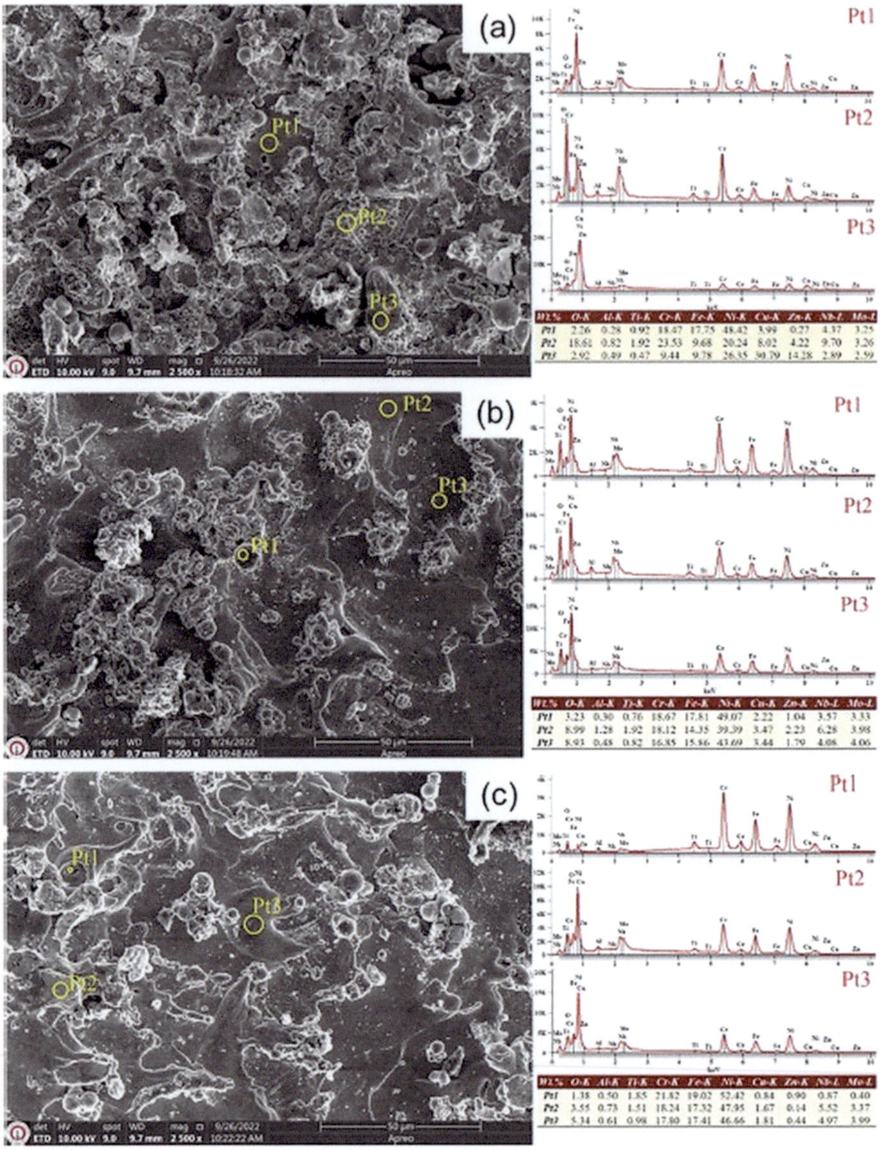

Fig. 1.8 SEM and EDS analysis after **a** rough, **b** semifinish, and **c** finish cut [53]

two and three passes, respectively. All machining operations were performed in pure water as the dielectric medium. Surface contamination and morphology were analyzed using Scanning Electron Microscopy (SEM) and Energy Dispersive X-ray Spectroscopy (EDS), as shown in Fig. 1.8. The results demonstrated that the micro-EDM process, due to its noncontact material removal mechanism, produced relatively

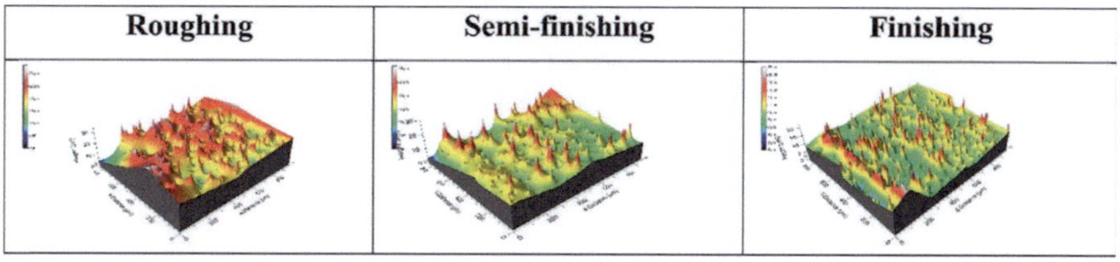

Fig. 1.9 Surface topographies after rough, semifinish, and finish cut [53]

clean surfaces. Only trace amounts of copper were detected on the machined surfaces, indicating minimal electrode material transfer. However, irregular crater formations were observed, which were attributed to the localized melting and resolidification of the material during discharge. These thermal effects increased surface roughness, particularly in rough machining. Nonetheless, the application of multiple machining passes effectively reduced surface roughness as shown in Fig. 1.9, by progressively refining the surface and mitigating the effects of remelted regions.

1.3.1.4 Hybrid Micromachining

Hybrid micromachining is a combination of multiple machining processes aimed at improving material removal rates, tool life, surface integrity, and machining efficiency, particularly for difficult-to-machine materials like ceramics, hard steels, and titanium alloys [54]. This process can be divided into assisted and combined techniques as presented in Fig. 1.10. Assisted techniques use external energy sources, such as ultrasonic vibrations, laser, or magnetic fields, to enhance machining performance. For instance, laser-assisted micromachining can preheat the workpiece, reducing cutting forces and improving tool life. Vibration-assisted micromachining, meanwhile, helps reduce process forces, enhances chip removal, and improves surface quality. These assisted processes are particularly useful for complex geometries that require submicron accuracy and a good surface finish.

Combined hybrid micromachining involves integrating different machining mechanisms that work simultaneously to achieve better process results. For example, laser-assisted jet electrochemical micromachining (LAJECM) combines the precision of laser localization with the material removal capability of electrochemical machining, resulting in improved dimensional accuracy and higher material removal rates. Another example is the integration of ultrasonic vibrations in micro-EDM, which improves flushing efficiency, increases material removal rates, and reduces heat-affected zones. Hybrid processes are essential for achieving high precision and efficiency when manufacturing microcomponents used in industries like automotive, aerospace, and medical instruments [55, 56].

Worts et al. [57] investigated femtosecond laser micromachining of Ti–6Al–4V components fabricated by LPBF. The LPBF process was conducted using a layer thickness of 30 μm, a laser power of 200 W, and a wavelength of 1040 nm to produce

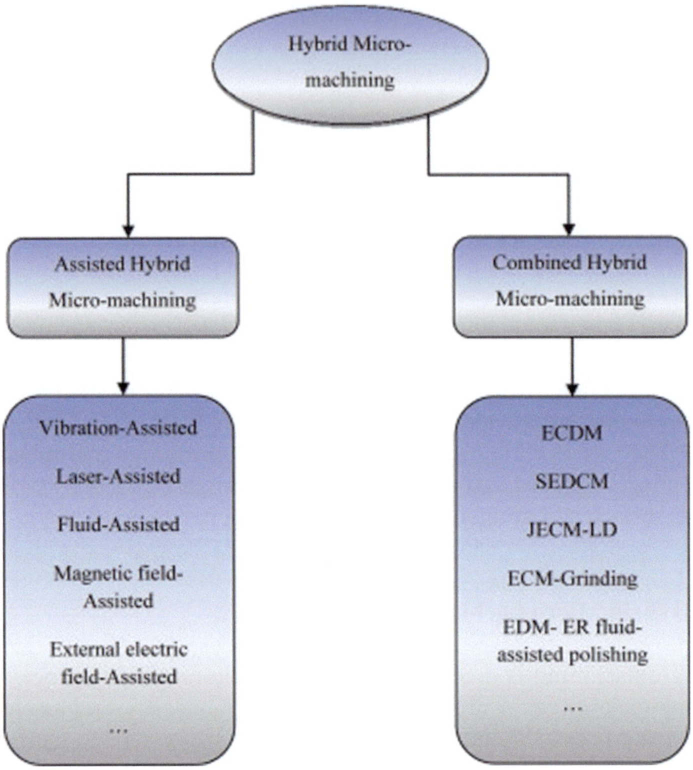

Fig. 1.10 Classification of hybrid micromachining processes [55]

specimens with dimensions of 250 mm × 250 mm × 300 mm. Subsequently, the specimens underwent femtosecond laser micromachining using a focal spot size of 16 μm. Surface analyses revealed that the average surface roughness of the as-built AM part was 4.23 μm, which was significantly reduced to 0.82 μm after femtosecond laser micromachining. Figure 1.11 presents the microstructural views and surface roughness profiles of the AM component before and after femtosecond laser treatment. A similar study was conducted by Ghosh et al. [58]. In their study, femtosecond laser micromachining operation was applied to SS316L that was produced by LPBF. As shown in Fig. 1.12, femtosecond laser micromachining is highly effective in improving the surface finish of AM parts, which often require postprocessing to mitigate their inherently high surface roughness.

Zheng et al. [59] investigated picosecond laser micromachining for AM components produced from SS316L powder. A hybrid manufacturing system integrating both additive and subtractive processes was developed for the study. The system incorporated two separate optical paths: one for continuous-wave (CW) laser AM and another for pulsed-laser subtractive micromachining. For the additive process, a 500 W continuous infrared laser with a wavelength of 1070 nm was employed. The CW

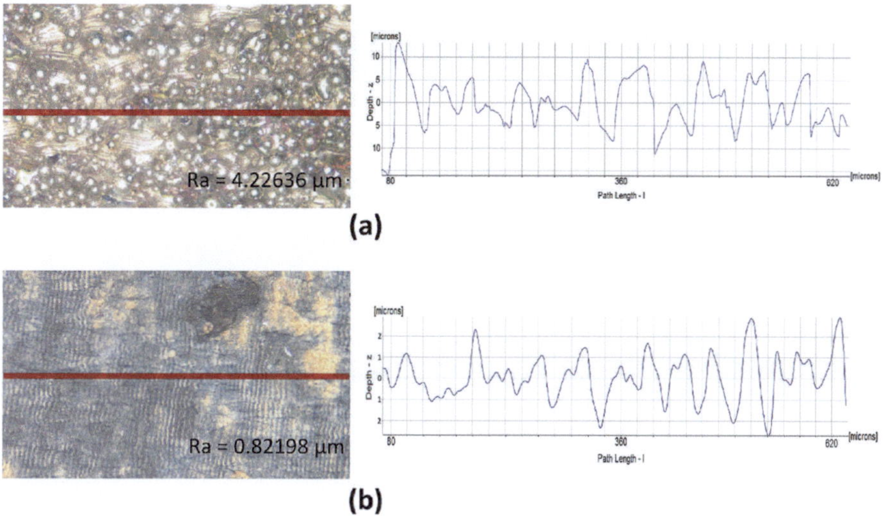

Fig. 1.11 Microstructural view and surface roughness profile for the AM part **a** before and **b** after femtosecond laser micromachining [57]

laser beam was collimated, directed through a scanning oscillator, and focused onto the build surface using an f-θ lens. For the subtractive process, a 50 W picosecond laser with a wavelength of 1030 nm and a pulse duration of 1.5 ps was used as the ultrashort pulse (USP) laser. Due to its lower beam quality, the USP laser beam was expanded to achieve a smaller focused spot size. Although the same f-θ lens was used for both laser paths, the slight difference in wavelengths led to noncoincident focal planes. To address this, the distance between the f-θ lens and the processing surface was finely adjusted to align the focal planes of both lasers, ensuring accurate and effective hybrid manufacturing. Figure 1.13 shows the experimental details of the hybrid system.

According to surface measurements, the average roughness of the machined part was significantly reduced from 12.10 to 1.65 μm. This demonstrates that the inherently rough surface profile of AM components can be substantially smoothed, thereby enhancing the overall surface quality through picosecond laser micromachining. Figure 1.14 presents the surface topography and roughness profiles of the AM parts before and after the micromachining process. Beyond surface finish improvements, picosecond laser micromachining also proves effective in enhancing the mechanical properties of AM products, particularly surface hardness. As shown in Fig. 1.15, the microhardness of the as-built AM specimen was measured at 184.6 HV. After picosecond laser process, the microhardness increased to values ranging from 215.3 to 275.8 HV, depending on the number of laser passes applied to the surface.

Bai et al. [60] investigated ultrasonic vibration-assisted machining (UEVAM) of AlSiMg0.75 alloys fabricated using the SLM technique. A range of process parameters (laser power, scanning speed, and hatch distance) were varied in the SLM

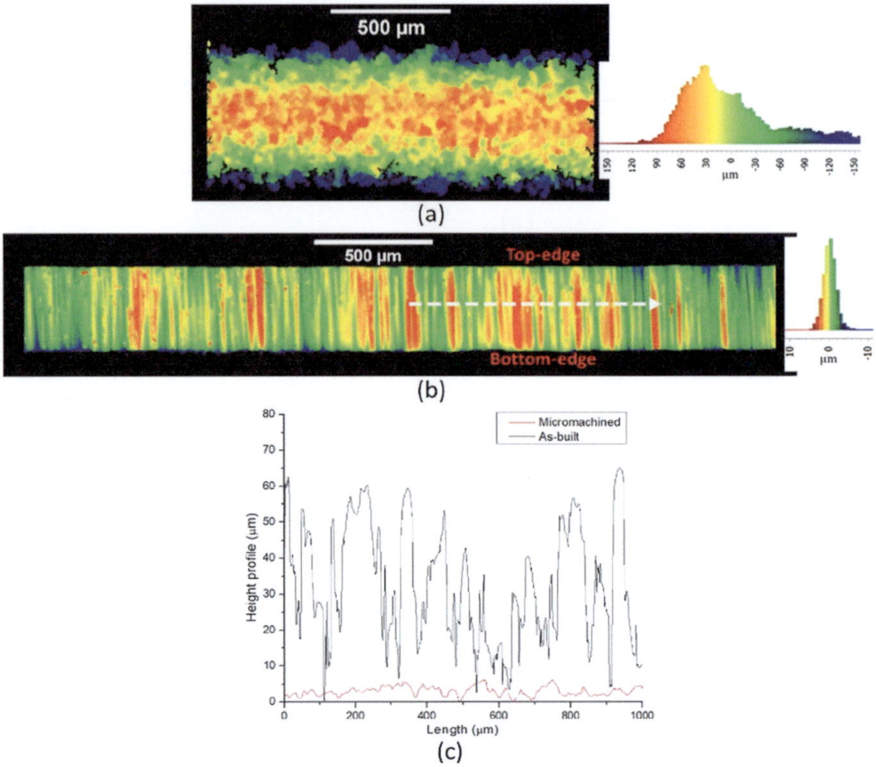

Fig. 1.12 2D surface profile of the AM part **a** before and **b** after femtosecond laser micromachining, and **c** comparison of surface roughness profiles [58]

process. The specimens were divided into two groups: as-built (AB) and heat-treated (HT), with the latter undergoing a heat treatment at 300 °C for 3 h followed by furnace cooling. Subsequently, turning operations were performed using ultrasonic vibrations applied to the cutting tool. The UEVAM process parameters included a spindle speed of 1500 rpm, a feed rate of 2 µm/rev, a depth of cut of 3 µm, a cutting speed of 20 mm/min, and a tool vibration frequency of 38.87 kHz. Optimal SLM parameters of 180 W laser power, 700 mm/s scanning speed, and 0.10 mm hatch distance yielded a maximum relative density of 99.6%. Surface roughness was evaluated via microstructural imaging and roughness profile measurements. As shown in Fig. 1.16, UEVAM significantly reduced surface roughness for both AB and HT conditions. For AB specimens, the average surface roughness (Ra) decreased by 53.76% and the maximum surface profile variation (Rz) decreased by 53.40% when transitioning from conventional turning (CON) to UEVAM. Similar improvements were observed in HT specimens, with reductions of 51.34% in Ra and 47.09% in

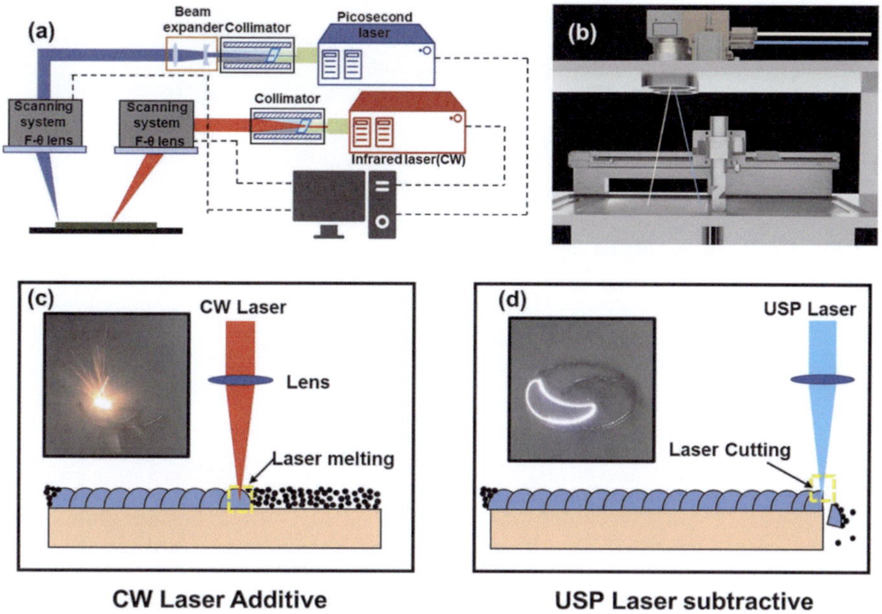

Fig. 1.13 Hybrid manufacturing system: **a** optical path system, **b** experimental setup, **c** additive process, and **d** subtractive process [59]

Rz. Figure 1.17 shows micrographs of AM parts after conventional and ultrasonic-assisted machining, highlighting the elimination of scratches and tool marks due to the application of ultrasonic vibrations.

1.4 Benefits and Challenges in Micromachining of AM Metallic Parts

Micromachining offers several key benefits when applied as a postprocessing technique for AM parts. One of the primary advantages of micromachining is its ability to significantly improve the surface finish of AM parts. Due to the layer-by-layer nature of AM processes, parts often exhibit high surface roughness and irregularities. Micromachining techniques such as micromilling, microdrilling, and micro-EDM are capable of smoothing rough surfaces, removing asperities, and achieving the fine tolerances required for critical applications. By refining the surface texture, micromachining enhances not only the aesthetic appearance of AM parts but also their functional performance, reducing friction, wear, and other surface-related issues [56]. On the other hand, AM processes can sometimes result in parts with dimensional inaccuracies due to issues such as thermal distortion, shrinkage, and misalignment. Micromachining allows for precise control over the final geometry of the

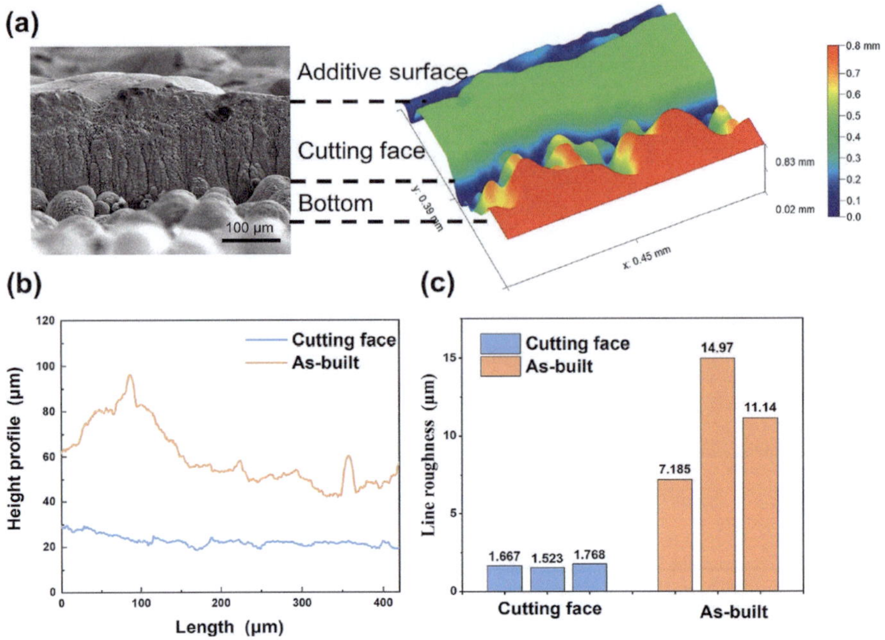

Fig. 1.14 **a** Sidewall view of the part after picosecond laser micromachining, **b** surface profile before and after processing, and **c** line roughness values [59]

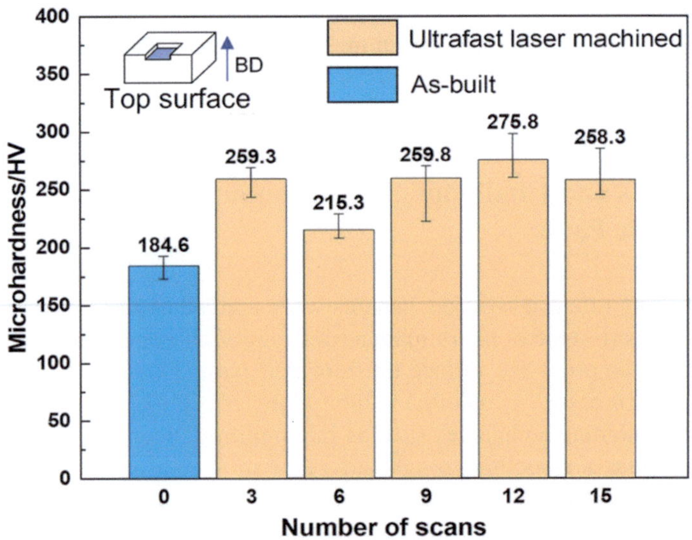

Fig. 1.15 Microhardness values of the specimens [59]

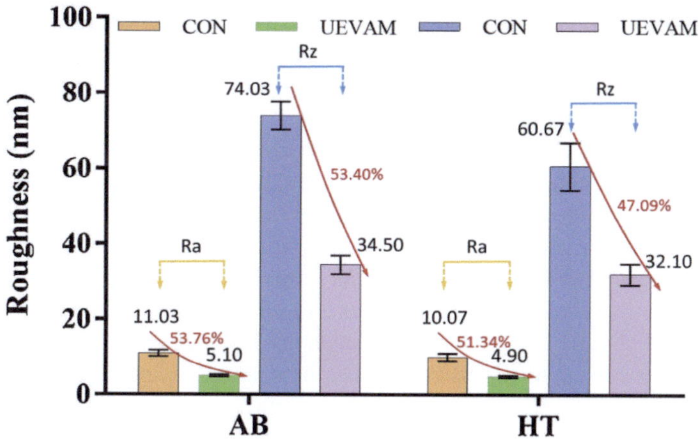

Fig. 1.16 Surface roughness results for AB and HT conditioned specimens [60]

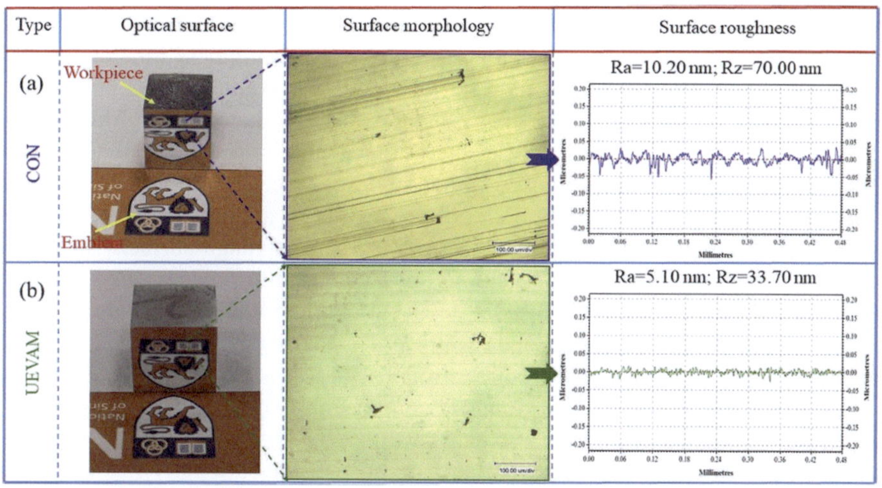

Fig. 1.17 Microviews of the specimens after **a** conventional turning and **b** ultrasonic vibration-assisted machining [60]

part, ensuring that it meets tight tolerance requirements. This is especially important for components that require high precision, such as those used in aerospace and medical applications. Micromachining can correct geometrical imperfections, bring parts into compliance with design specifications, and ensure proper fit and functionality [61, 62]. In addition, AM parts often suffer from microstructural defects, including porosity, anisotropy, and residual stresses, which can negatively affect their mechanical properties. Micromachining can help mitigate these defects by removing material from the surface or specific areas where such issues are most prominent. For example, micromilling and micro-EDM can be used to refine areas with localized defects or inconsistencies, improving the overall material integrity and reliability of the part. This is particularly important in applications where the performance and longevity of the part are critical [63, 64].

Micromachining AM parts presents distinct challenges compared to conventionally manufactured parts, primarily due to the unique characteristics inherent to AM processes. These differences impact factors such as machinability, tool wear, surface quality, and dimensional accuracy. One of the primary distinctions is related to the microstructure of AM materials. AM parts exhibit anisotropy and porosity, while conventionally manufactured materials tend to have a more uniform and homogeneous microstructure. In AM parts, the layer-by-layer construction process introduces internal stresses, microstructural variations, and the potential for defects such as voids and lack of fusion, which are less common in conventionally produced materials. These irregularities in AM parts can cause unpredictable cutting forces and tool wear during micromachining, making it more challenging to achieve consistent results. In terms of tool wear, AM materials, especially those made from hard metals like Inconel and titanium alloys, tend to exhibit higher levels of tool wear compared to their conventionally manufactured counterparts. This is due to the presence of residual stresses, hard phases, and surface roughness inherent in AM processes, which increase the abrasion and thermal stress on cutting tools. In contrast, conventionally manufactured parts, particularly those produced through forging or casting, have more controlled microstructures that generally result in less aggressive tool wear during micromachining. Another key difference lies in surface integrity. AM parts typically require additional postprocessing to achieve the same level of surface finish as conventionally manufactured parts. The rough surfaces and stair-stepping effects seen in AM parts due to the layered build process often result in poorer initial surface quality. Conversely, conventionally produced parts are typically more refined after manufacturing and require less intensive postprocessing to achieve the desired surface finish. Finally, dimensional accuracy is another area where AM parts differ from conventionally produced components. The thermal cycles in AM, combined with residual stresses, can cause warping or dimensional distortions that require more extensive micromachining to correct. Conventionally manufactured parts generally have better initial dimensional accuracy, requiring less correction during postprocessing [32, 56].

Minimum quantity lubrication (MQL) is particularly beneficial in addressing these challenges, because it effectively reduces cutting forces and built-up edges, which are common in machining difficult-to-cut metallic AM materials. By applying fine sprays of lubricant in precise quantities, MQL creates a thin lubrication layer that reduces friction between the cutting tool and workpiece, leading to lower tool wear. This is especially important for AM parts, where surface roughness and hard phases can intensify abrasion on the cutting tool. Furthermore, MQL's ability to reduce cutting forces and temperatures at the tool–workpiece interface helps prevent tool degradation, especially in materials with poor thermal conductivity, such as titanium. The fine sprays and constant or high-frequency lubricant delivery in MQL also improve surface finish by minimizing burr formation and enhancing chip evacuation. Thus, implementing MQL can significantly improve the surface integrity and tool life during micromachining of AM parts [63, 64].

Tool wear and surface integrity issues in micromachining are often deteriorated by the hard and abrasive nature of AM materials, as well as the thermal and mechanical stresses involved. Selecting appropriate coatings for microtools is crucial for mitigating these problems. For example, TiN and AlCrTiN coatings are excellent for machining hard steels, which are often used in AM processes. These coatings provide superior hardness and wear resistance, which prolongs the tool life when dealing with the high cutting forces and abrasive surfaces of AM parts. For stainless steel and titanium alloys, coatings like TiAlN and diamond-like carbon (DLC) are recommended. DLC, in particular, offers excellent thermal stability and low friction, which is critical when machining titanium alloys that tend to generate excessive heat during cutting. The proper use of these coatings helps reduce tool wear and built-up edges, ensuring a more consistent surface finish and reducing the likelihood of tool failure during micromachining of AM parts [63, 64].

1.5 Conclusions

Metallic materials have been widely produced by AM techniques for various sectors, such as aerospace, automotive, and medical industries. AM products provide a wide range of advantages for the applications like topology optimization, lightweight design, near-net-shape production, etc. However, metallic products fabricated by AM generally require postprocessing operations to finalize the component for service conditions. AM is characterized by its layer-by-layer material deposition process, and often leads to components with suboptimal surface finishes. These parts typically display elevated levels of surface roughness and geometrical imperfections when compared to those produced using conventional manufacturing techniques. These deficiencies can pose significant challenges in meeting the tight dimensional tolerances and surface smoothness required in high-precision applications. To address these issues, micromachining plays a crucial postprocessing role. It is a high-precision material removal technique that enables the refinement of AM-produced surfaces, helping to eliminate surface irregularities and achieve the exact specifications demanded by various industries. Moreover, parts manufactured through AM frequently exhibit anisotropic microstructures as well as dimensional inconsistencies due to the nature of the additive process. These factors can compromise the mechanical performance, reliability, and longevity of the components. By employing micromachining, manufacturers can precisely remove excess or undesired material, enhance surface integrity, and ensure tighter dimensional control. This significantly improves the functional characteristics of the final product, making it more reliable and performance-ready. As a result, micromachining is not only essential for improving the visual appeal and mechanical robustness of AM parts but also serves a critical function in qualifying these components for demanding and safety–critical applications. Industries such as aerospace, medical device manufacturing, and automotive engineering, where component reliability and performance are paramount, particularly benefit from this postprocessing technique to ensure that AM parts meet rigorous operational standards.

This chapter presents an overview of micromachining processes applied to metallic parts produced by AM. It begins by outlining the AM technologies commonly used for metallic materials, followed by a detailed discussion of specific processes. Furthermore, micromachining techniques suitable for these materials are discussed, including micromilling, microdrilling, micro-EDM, and hybrid micromachining methods. Process mechanisms and outcomes reported in previous studies are discussed to highlight the role of micromachining as a critical postprocessing step. In particular, dimensional accuracy and surface quality, which are mostly challenging in AM metallic components, can be significantly improved through micromachining. Given this potential, process optimization emerges as a key strategy to achieve enhanced performance and reliability in AM-produced metallic parts.

References

1. Bourhis FL, Kerbrat O, Dembinski L, Hascoet JY, Mognol P. Predictive model for environmental assessment in additive manufacturing process. Procedia CIRP. 2014;15:26–31.
2. Huang SH, Liu P, Mokasdar A, Hou L. Additive manufacturing and its societal impact: a literature review. Int J Adv Manuf Technol. 2013;67(5–8):1191–1203.
3. Ford S, Despeisse M. Additive manufacturing and sustainability: an exploratory study of the advantages and challenges. J Clean Prod. 2016;137:1573–1587.
4. Arrizubieta JI, Ukar O, Ostolaza M, Mugica A. Study of the environmental implications of using metal powder in additive manufacturing and its handling. Metals. 2020;10(2):261.
5. Murr LE, Martinez E, Amato KN, Gaytan SM, Hernandez J, Ramirez DA, et al. Fabrication of metal and alloy components by additive manufacturing: examples of 3D materials science. J Mater Res Technol. 2012;1(1):42–54.
6. Renjith SC, Park K, Okudan Kremer GE. A design framework for additive manufacturing: integration of additive manufacturing capabilities in the early design process. Int J Precis Eng Manuf. 2020;21:329–345.
7. Thompson MK, Moroni G, Vaneker T, Fadel G, Campbell RI, Gibson I, et al. Design for additive manufacturing: trends, opportunities, considerations, and constraints. CIRP Ann. 2016;65(2):737–760.
8. Kumar SP. Measurement and uncertainty analysis of surface roughness and material removal rate in micro turning operation and process parameters optimization. Measurement. 2019;140:538–547.
9. Maleki E, Bagherifard S, Bandini M, Guagliano M. Surface post-treatments for metal additive manufacturing: progress, challenges, and opportunities. Addit Manuf. 2020;32:101619.
10. Yang S, Min W, Ghibaudo J, Zhao YF. Understanding the sustainability potential of part consolidation design supported by additive manufacturing. J Clean Prod. 2019;232:722–738.
11. Liu ZY, Li C, Fang XY, Guo YB. Energy consumption in additive manufacturing of metal parts. Procedia Manuf. 2018;26:834–845.
12. Hassanin H, Elshaer A, Benhadj-Djilali R, Modica F, Fassi I. Surface finish improvement of additive manufactured metal parts. In: Davim JP, editor. Surface Integrity in Machining. Berlin: Springer; 2017. p. 145–164.

13. DebRoy T, Wei HL, Zuback JS, Mukherjee T, Elmer JW, Milewski JO, et al. Additive manufacturing of metallic components—process, structure and properties. Prog Mater Sci. 2018;92:112–224.
14. Greco S, Kieren-Ehses S, Kirsch B, Aurich JC. Micro milling of additively manufactured AISI 316L: impact of the layerwise microstructure on the process results. Int J Adv Manuf Technol. 2021;112(1–2):361–373.
15. Pradeep PV, Paul A, Paul L, Davim JP. Application of micro machining and additive manufacturing processes for difficult-to-cut materials in aerospace industry—a review. In: Paul L, Davim JP, editors. Modern Manufacturing Processes. Amsterdam: Elsevier; 2020. p. 227–262.
16. Thijs L, Verhaeghe F, Craeghs T, Humbeeck JV, Kruth JP. A study of the microstructural evolution during selective laser melting of Ti–6Al–4V. Acta Mater. 2010;58(9):3303–3312.
17. Leuders S, Lieneke T, Lammers S, Tröster T, Niendorf T. On the mechanical behaviour of titanium alloy TiAl6V4 manufactured by selective laser melting: fatigue resistance and crack growth performance. Int J Fatigue. 2013;48:300–307.
18. Rashid R, Masood SH, Ruan D, Palanisamy S, Rahman Rashid B, Xu W, et al. Effect of energy per layer on the anisotropy of selective laser melted AlSi12 aluminium alloy. Addit Manuf. 2018;22:426–439.
19. Ansari P, Rehman AU, Pitir F, Veziroglu S, Mishra YK, Aktas OC, Salamci MU. Selective Laser Melting of 316L Austenitic Stainless Steel: Detailed Process Understanding Using Multiphysics Simulation and Experimentation. Metals. 2021;11(7):1076.
20. Rawal S, Brantley J, Karabudak N. Additive manufacturing of Ti-6Al-4V alloy components for spacecraft applications. In: Proceedings of the 6th International Conference on Recent Advances in Space Technologies (RAST). Istanbul: IEEE; 2013. p. 5–8.
21. Mohammad A, Abdel-Hamid M, Nasr M, Kovacs Z, Pimenov DY, El Rayes MM. Effect of energy input on microstructure and mechanical properties of titanium aluminide alloy fabricated by the additive manufacturing process of electron beam melting. Materials (Basel). 2017;10(2):211.
22. Zhang LC, Liu Y, Li SJ, Hao YL. Additive manufacturing of titanium alloys by electron beam melting: a review. Adv Eng Mater. 2018;20(5):1700842.
23. Rafi HK, Karthik NV, Gong H, Starr TL, Stucker BE. Microstructures and mechanical properties of Ti6Al4V parts fabricated by selective laser melting and electron beam melting. J Mater Eng Perform. 2013;22(12):3872–3883.
24. Gruber H., Henriksson M., Hryha E, Nyborg L. Effect of Powder Recycling in Electron Beam Melting on the Surface Chemistry of Alloy 718 Powder. Metall Mater Trans A. 2019;50:4410–4422.
25. Tekumalla S, Tosi R, Tan X, Seita M. Directed energy deposition and characterization of high-speed steels with high vanadium content. Addit Manuf Lett. 2022;2:100029.
26. Ahn DG. Direct metal additive manufacturing processes and their sustainable applications for green technology: a review. Int J Precis Eng Manuf Green Technol. 2016;3:381–395.
27. Ahn DG. Directed energy deposition (DED) process: state of the art. Int J Precis Eng Manuf Green Technol. 2021;8:703–742.
28. Kumbhar NN, Mulay AV. Post processing methods used to improve surface finish of products which are manufactured by additive manufacturing technologies: a review. J Inst Eng India Ser C. 2018;99(4):481–487.
29. Ojo OO, Taban E. Post-processing treatments–microstructure–performance interrelationship of metal additive manufactured aerospace alloys: a review. Mater Sci Technol. 2023;39(1):1–41.
30. Tyagi P, Goulet T, Riso C, Garcia-Moreno F. Reducing surface roughness by chemical polishing of additively manufactured 3D printed 316 stainless steel components. Int J Adv Manuf Technol. 2019;100(9–12):2895–2900.
31. Duan X, Su F, Gao S, Zhu K, Deng B, Zhang Y. Relationship between energy efficiency and surface morphologies in micro-milling of SLM Inconel 718. Journal of Materials Research and Technology2024; 31:1473–82.
32. Balázs BZ, Geier N, Takács M, Kundrák J, Markopoulos AP, Ghadbeigi H, et al. A review on micro-milling: recent advances and future trends. Int J Adv Manuf Technol. 2021;112:655–684.

33. Chen Z, Wu X, Ke W, Shen J, Jiang F, Zhu L, et al. Tool wear mechanisms of PCD micro end mill in machining of additive manufactured titanium alloy. Int J Adv Manuf Technol. 2023;127(7–8):3269–3280.
34. de Assis CLF, Mecelis GR, Coelho RT. An investigation of stainless steel 316L parts produced by powder bed fusion submitted to micro-endmilling operations. Int J Adv Manuf Technol. 2020;109(7–8):1867–1880.
35. Sandeep K, Paolo P, Salvatore C, Massimiliano A. Micromilling of metallic feedstock produced by extrusion additive manufacturing. In: Proceedings of the WCMNM 2018 World Congress on Micro and Nano Manufacturing; 2018. p. 978–981.
36. Abeni A, Ginestra PS, Attanasio A. Micro-milling of selective laser melted stainless steel. In: Davim JP, editor. Machining of Stainless Steels and Super Alloys. Cham: Springer; 2021. p. 1–12.
37. Allegri G, Colpani A, Ginestra PS, Attanasio A. An experimental study on micro-milling of a medical grade Co-Cr-Mo alloy produced by selective laser melting. Materials (Basel). 2019;12(13):2208.
38. Du H, Wu C, Li D, Yip WS, Wang Z, To S. Feasibility study on ultraprecision micro-milling of the additively manufactured NiTi alloy for generating microstructure arrays. J Mater Res Technol. 2023;25:55–67.
39. Gao Z, Nowak JF, Corbin J, Delgado L, Samuel J. Sensitivity of micromilling responses to grain variations in wire arc additively manufactured aluminum alloy 4043. J Micro Nano-Manuf. 2020;8(4):041001.
40. Ji H, Gupta MK, Song Q, Cai W, Zheng T, Zhao Y, et al. Microstructure and machinability evaluation in micro milling of selective laser melted Inconel 718 alloy. J Mater Res Technol. 2021;14:348–362.
41. Sadiq MA, Hoang NM, Valencia N, Obeidat S, Hung WNP. Experimental study of micromilling selective laser melted Inconel 718 superalloy. Procedia Manuf. 2018;26:983–992.
42. Abeni A, Loda D, Özel T, Attanasio A. Analytical force modelling for micro milling additively fabricated Inconel 625. Prod Eng Res Dev. 2020;14(5–6):613–627.
43. Yu T, Zhu X, Yu H, Wu P, Li C, Han X, et al. Material extrusion-based additive manufacturing of zirconia toughened alumina: machinability, mechanical properties and biocompatibility. J Manuf Process. 2023;94:120–132.
44. Azim S, Gangopadhyay S, Siba Sankar Mahapatra, Mittal RK. Performance evaluation of CrAlN and TiAlN coatings deposited by HiPIMS in micro drilling of a Ni-based superalloy. Surface and Coatings Technology. 2022;449:128980–0.
45. Cheng K, Huo D, editors. Micro-cutting: fundamentals and applications. Hoboken: Wiley; 2013.
46. Sorgato M, Zanini F, Bertolini R, Ghiotti A, Bruschi S. Improvement of micro-hole precision by ultrasound-assisted drilling of laser powder bed fused Ti6Al4V titanium alloy. Precis Eng. 2020;66:31–41.
47. Rysava Z, Bruschi S, Carmignato S, Medeossi F, Savio E, Zanini F. Micro-drilling and threading of the Ti6Al4V titanium alloy produced through additive manufacturing. Procedia CIRP. 2016;46:583–586.
48. Godara M, Joshy J, Kuriachen B, Mathew J. Experimental analysis on the effect of build directions during microdrilling of SLM fabricated maraging steel (18-Ni-300). Prog Addit Manuf. 2023.
49. Joshy J, Kuriachen B. Influence of post-processing and build orientation on the micro-machinability and chip formation during micro-drilling of L-PBF AlSi10Mg. CIRP J Manuf Sci Technol. 2023;45:35–48.
50. Madhavi M, Srikanth P. Characterization of process parameters in micro EDM on stainless steel. Materials Today Proceedings. 2023;90:164–8.
51. Jackson MJ, Davim JP, editors. Micro and nano machining of engineering materials. Berlin: Springer; 2010.
52. Ozaner OC, Dursun G, Akbulut G. Effects of wire-EDM parameters on the surface integrity and mechanical characteristics of additively manufactured Inconel 939. Mater Today Proc. 2021;38:1861–1865.

53. Gokcekaya O, Günen A, Ceritbinmez F, Bahador A, Nakano T, Çetin M. Wire-EDM performance and surface integrity of Inconel 718 with unique microstructural features fabricated by laser powder bed fusion. Int J Adv Manuf Technol. 2024;130(9–10):4513–4528.
54. Kibria G, Sarkar BR, Pradhan BB, Bhattacharyya B. Comparative study of different dielectrics for micro-EDM performance during microhole machining of Ti-6Al-4V alloy. Int J Adv Manuf Technol. 2010;48:557–570.
55. Chavoshi SZ, Luo X. Hybrid micro-machining processes: A review. Precision Engineering. 2015;41:1–23.
56. Manikandan P, Venkatesan K. Advancements in micromachining of additive manufactured materials: a comprehensive review. Mater Manuf Process. 2024;39(11):1469–1520.
57. Worts, N., Jones, J., Squier, J. Surface Structure Modification of Additively Manufactured Titanium Components via Femtosecond Laser Micromachining. Opt. Commun. 2019;430:352–357.
58. Ghosh, A., Biswas, S., Turner, T., Kietzig, A.M., Brochu, M.S. Surface, Microstructure, and Tensile Deformation Characterization of LPBF SS316L Microstruts Micromachined with Femtosecond Laser. Mater. Des. 2021;210:110045.
59. Zheng, B., Trofimov, V., Yang, Y., Liu, L., Feng, Y., Zheng, Z., et al. Study on Additive and Subtractive Manufacturing of High-Quality Surface Parts Enabled by Picosecond Laser. J. Mater. Process. Technol. 2023;318:118013.
60. Bai, Y., Shi, Z., Lee, Y.J., Wang, H. Optical Surface Generation on Additively Manufactured AlSiMg0.75 Alloys with Ultrasonic Vibration-Assisted Machining. J. Mater. Process. Technol. 2020;280:116597.
61. Wang, H., Hu, Y., Ning, F., Cong, W. Ultrasonic Vibration-Assisted Laser Engineered Net Shaping of Inconel 718 Parts: Effects of Ultrasonic Frequency on Microstructural and Mechanical Properties. J. Mater. Process. Technol. 2020;276:116395.
62. Gomes MC, Gonçalves A, de Oliveira D, Figueiredo GV, Barragán SK, Machado AR, et al. Micro-machining of additively manufactured metals: a review. Int J Adv Manuf Technol. 2021;118(7–8):2059–2078.
63. Li G, Chandra S, Rahman A, Palanisamy S, Ding S. Machinability of additively manufactured titanium alloys: a comprehensive review. J Manuf Process. 2022;75:72–99.
64. Sebbe NPV, Fernandes F, Sousa FC. Hybrid manufacturing processes used in the production of complex parts: a comprehensive review. Metals (Basel). 2022;12(11):1874.

Chapter 2
Surface Micro-topography in Ultrasonic-Assisted Machining: Medical and Industrial Applications

Mohammad Lotfi

Abstract This chapter presents an overview of the significance of surface micro-topography (SMT) in ultrasonic-assisted machining (UAM) processes. It explains the fundamental principles of ultrasonic vibrations, specifically linear, elliptical, and three-dimensional modes, and how they are effectively applied in various machining techniques such as turning, drilling, and milling. Key parameters related to SMT, such as contact angle, wettability, and anisotropic wettability, are discussed in the context of characterizing and evaluating surface quality and behavior. The wide range of SMT applications in both industrial manufacturing and medical fields is also discussed, with an emphasis on how surface textures influence functional performance.

Keywords Micromanufacturing · Ultrasonic-assisted machining · Material removal · Surface micro-topography

2.1 Introduction

In today's highly competitive manufacturing landscape, the demand for products with superior quality and precision has become increasingly critical. Consequently, there is a growing inclination among industrial professionals toward adopting advanced manufacturing processes. Surface micro-topography (SMT) has emerged as a significant focus area in both manufacturing and biomedical engineering, as the quality of SMT directly influences key properties such as wear resistance, fatigue life, and surface wettability of machined components [1]. The main parameters affecting SMT include the workpiece material, the geometry of the microtexture, and surface integrity [2]. This topic gains even more importance when considering the machining of advanced alloys and the challenges associated with their production.

M. Lotfi (✉)
Student Research Committee, Abadan University of Medical Sciences, Abadan, Iran
e-mail: lotfi.m26@gmail.com

School of Mechanical Engineering, Sharif University of Technology, Tehran, Iran

S. Gürgen (ed.), *Micromanufacturing Operations*,
https://doi.org/10.1007/978-3-032-04193-7_2

Accordingly, advanced alloys, such as nickel-based superalloys and titanium, are commonly employed in specialized applications due to their exceptional properties, including high resistance to fatigue and corrosion, as well as the ability to retain strength and hardness at elevated temperatures [3]. While these characteristics make such materials ideal for manufacturing advanced components in both medical and industrial sectors, they also present significant challenges during the machining process [4]. Machining operations, which remain among the most widely utilized methods for manufacturing mechanical components, continue to encounter difficulties when processing advanced alloys [5]. To address these challenges, several techniques have been developed to enhance the machining process, including cryogenic machining, minimum quantity lubrication (MQL), and ultrasonic-assisted machining (UAM). Among these, UAM has been demonstrated to be particularly effective in improving SMT when machining advanced materials [6].

This chapter provides an investigation into the role of SMT within UAM techniques. It begins by outlining the fundamental mechanisms behind ultrasonic vibrations—covering linear, elliptical, and three-dimensional modes—and their implementation in various machining processes such as turning, drilling, and milling. Following this, the chapter discusses the critical parameters associated with SMT, including contact angle, wettability, and anisotropic wettability, highlighting their significance in assessing surface characteristics. Finally, the discussion covers the diverse applications of SMT across both industrial and medical fields, underscoring the practical benefits and improvements these surface modifications can provide in terms of surface, microstructural and mechanical properties as well as biological integration.

2.2 Ultrasonic-Assisted Machining

One of the most widely adopted techniques in modern manufacturing is the application of ultrasonic waves, and in recent years, significant research efforts have been devoted to integrating these waves into various processes. Ultrasonics refer to sound waves with frequencies above the audible range of human hearing. These are elastic waves capable of propagating through gaseous, liquid, and solid media. A key characteristic of ultrasonic waves is their strong directionality and ability to concentrate energy within a localized region. In solids, ultrasonic vibrations typically exhibit oscillation amplitudes on the order of several micrometers [7, 8]. One prominent application of ultrasonic vibrations is in the machining of materials.

UAM is classified as a nonconventional manufacturing process, particularly effective for machining advanced alloys. In this method, a relative motion is introduced between the cutting tool and the workpiece, typically in the form of high-frequency (approximately 20 kHz) vibrations with a small amplitude (around 10 μm) [9]. The tool vibration can be linear or elliptical and may occur in one, two, or three directions. Figure 2.1 shows a schematic illustration of tool vibration in UAM. In this figure, t_1, t_2, and φ represent the uncut chip thickness, deformed chip thickness, and shear

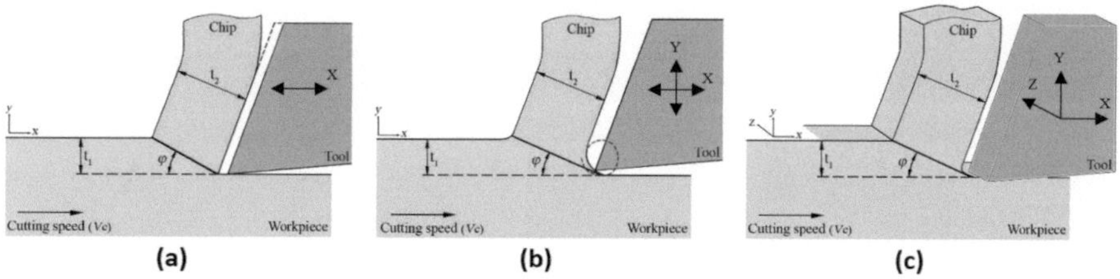

Fig. 2.1 Tool vibrations in UAM: **a** one, **b** two, and **c** three directions

angle, respectively. Figure 2.1a depicts linear vibration, where the cutting tool oscillates in the direction of the cutting speed (V_c). The instantaneous tool position (X_t) is given by Eq. 2.1, where ω is the angular frequency, a is the vibration amplitude, and t is time [10].

$$X_t = a\sin(\omega t) + V_c t \tag{2.1}$$

Figure 2.1b shows two-dimensional elliptical vibration. In this mode, an additional motion is superimposed on the cutting tool in the feed direction (Y_t). The vibration amplitude in the feed direction is denoted by b, and α represents the phase shift between the two oscillatory motions. The tool position under elliptical vibration is calculated using Eq. 2.2.

$$\left\{\begin{array}{c} X_t = a\sin(\omega t) + V_c t \\ Y_t = b\cos(\omega t + \alpha) \end{array}\right\} \tag{2.2}$$

Figure 2.1c shows three-dimensional elliptical vibration, with the corresponding tool position described by Eq. 2.3.

$$\left\{\begin{array}{c} X_t = a\sin(\omega t) + V_c t \\ Y_t = b\sin(\omega t + \alpha_1) \\ Z_t = c\sin(\omega t + \alpha_2) \end{array}\right\} \tag{2.3}$$

These types of tool vibrations are commonly employed in ultrasonic-assisted turning and boring operations, where the workpiece rotation determines the cutting speed and the feed motion is governed by the movement of the cutting tool [11].

A setup for a three-dimensional vibratory cutting tool is shown in Fig. 2.2. The configuration of the piezoelectric (PZT) elements and the geometry of the horn are key parameters in generating ultrasonic vibrations. As shown in the experimental setup, the type and number of PZTs excited determine whether the tool vibrates in one, two, or three dimensions. The PZT elements intermittently change thickness in response to applied electrical voltage, producing the desired vibrations. Moreover, a gap sensor is used to measure the vibration amplitude in various cutting directions, while an oscilloscope displays the amplitude values in real time.

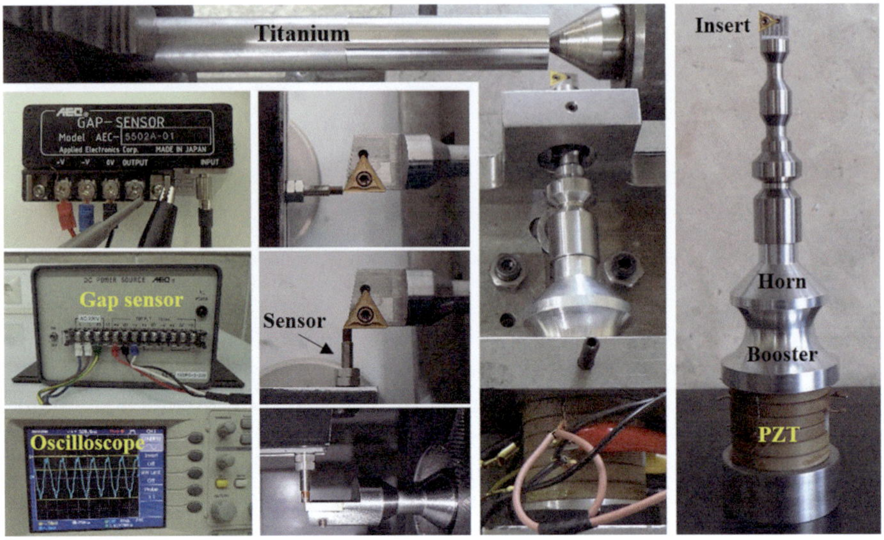

Fig. 2.2 An experimental setup for ultrasonic-assisted turning of titanium alloys [12]

Unlike turning and boring processes, in drilling and milling operations, the movements of the workpiece and cutting tool correspond to feed and rotational speeds, respectively [13, 14]. As illustrated in Fig. 2.3, the tool vibration can take various forms, including longitudinal vibration, longitudinal-torsional vibration, and elliptical vibration. Figure 2.4 shows a setup for a rotary ultrasonic-assisted drilling process [15]. This tool generates longitudinal and longitudinal-torsional vibrations in the feed direction. The horn determines the type of vibration produced. The PZTs and backing components are housed within the body of this vibratory rotary tool. To implement this setup, the computer numerical control (CNC) machine head must be replaced.

2.3 Surface Micro-topography

SMT is characterized by a series of peaks and valleys. It comprises several components, the most significant of which are surface roughness and texture geometry [16]. Given that SMT quality influences critical properties such as wear resistance, fatigue life, and surface wettability, considerable attention has been devoted to understanding the types of SMT that can enhance the performance and quality of manufactured components.

As previously noted, surface roughness is a key component of SMT. It is defined as the deviation of a real surface from its ideal form, measured in the direction of the surface normal. A surface with large deviations is considered rough, while one

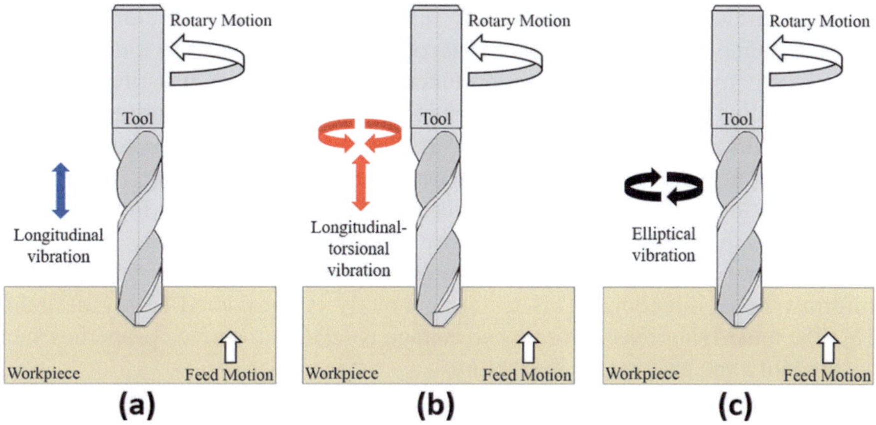

Fig. 2.3 Cutting tool vibration in drilling and milling operations: **a** longitudinal, **b** longitudinal-torsional, and **c** elliptical vibrations [15]

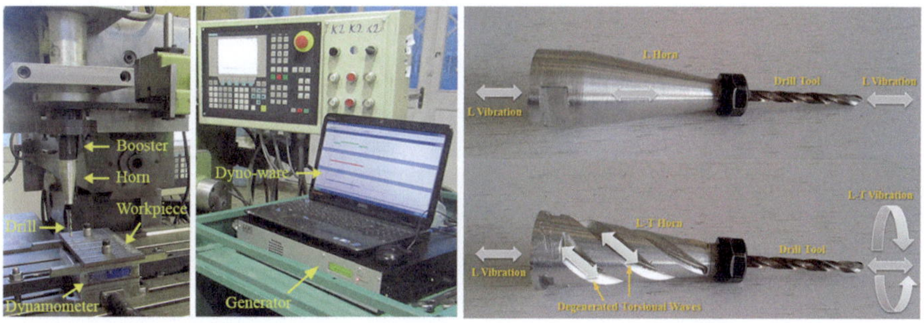

Fig. 2.4 A rotary ultrasonic-assisted drilling tool capable of generating longitudinal and longitudinal-torsional vibrations [15]

with minimal deviations is classified as smooth. In tribology, rough surfaces tend to wear more quickly and exhibit higher coefficients of friction compared to smooth surfaces [17]. On one hand, surface roughness is a critical factor in determining the service life of mechanical components, as surface irregularities often serve as initiation sites for cracks and mechanical failures. On the other hand, roughness can be advantageous in certain applications by enhancing the adhesion between surfaces or between a surface and another component. Conversely, reducing surface roughness may improve surface integrity but can diminish adhesion strength [18]. As a result, the optimal roughness level varies depending on the application. For example, in medical applications, surface adhesion is often prioritized over smoothness [19].

Surface texture geometry is another critical parameter of SMT. Beyond the general roughness or smoothness of a surface, the uniformity and regularity of peaks and valleys across different directions are also important. This is because a consistent surface texture ensures that the manufactured part exhibits uniform properties across its entire surface—that is, it maintains surface integrity [20]. Therefore, manufacturing methods must be capable of accounting for both surface roughness and texture geometry to ensure the desired functional performance of the final component. Accordingly, the desired surface characteristics cannot typically be achieved using conventional machining methods, as these processes often produce surface roughness in a single direction. For instance, in conventional turning, the resulting peaks and valleys align with the feed direction [21], as illustrated in Fig. 2.5a. This limitation can be overcome through the use of UAM. As shown in Fig. 2.5b, UAM produces a more isotropic SMT, where semispherical surface textures can be formed uniformly in all directions. This texture geometry is considered highly desirable [22]. The underlying reason for this advantage is related to surface properties such as wettability and anisotropic wettability.

Wettability refers to the tendency of a liquid to spread across a solid surface. It is commonly characterized by the contact angle formed between the liquid and the solid: as the contact angle decreases, wettability increases. Figure 2.6a schematically illustrates the ideal contact angle (θ_{Ideal}). As shown in the figure and described by Eq. 2.4, the ideal contact angle is calculated based on the maximum height (H) and the base radius (L) of the water droplet [23].

$$\theta_{\mathrm{Ideal}} = 2\arctan\left(\frac{H}{L}\right) \tag{2.4}$$

This equation applies to an ideal, perfectly smooth surface. However, conventionally machined surfaces exhibit surface roughness, as illustrated in Fig. 2.6b.

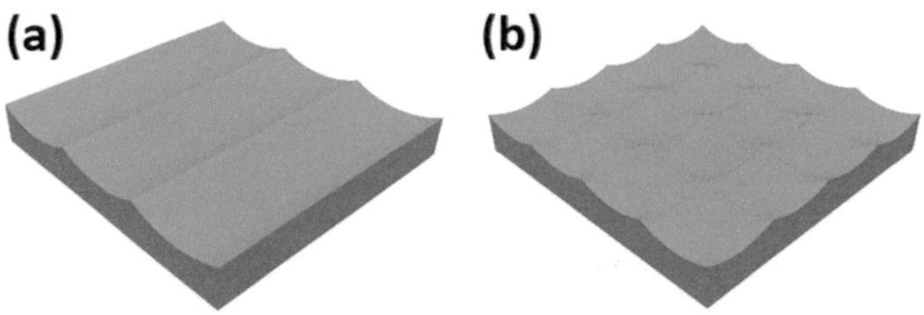

Fig. 2.5 A schematic illustration of SMT in **a** conventional machining and **b** UAM

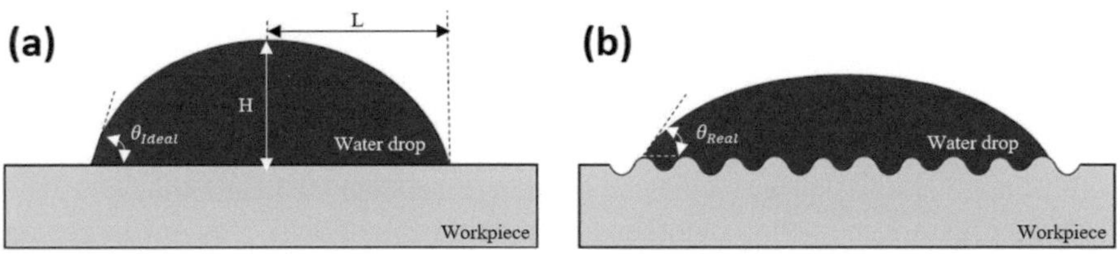

Fig. 2.6 A schematic illustration of **a** ideal and **b** real contact angle between water drop and workpiece

Therefore, to determine the real contact angle (θ_{Real}) on such surfaces, a modified form of the equation must be used. According to Eq. 2.5, this correction accounts for the surface roughness through a parameter known as the roughness ratio (R). The real contact angle is defined as the angle between the actual surface profile and the tangent to the liquid–vapor interface at the point of contact.

$$\cos\theta_{\text{Real}} = R\cos\theta_{\text{Ideal}} \tag{2.5}$$

Moreover, during UAM, microtextures are introduced onto the surface, further influencing wettability. As a result, the equation for calculating the real contact angle must be further modified by incorporating the developed interfacial area ratio, denoted as S_{dr}. Accordingly, Eqs. 2.6 and 2.7 are used to account for the effects of both surface roughness and microtexture on wettability.

$$S_{dr} = \frac{\text{Texture surface area} - \text{Cross sectional area}}{\text{Cross sectional area}} \tag{2.6}$$

$$R = 1 + \frac{S_{dr}}{100} \tag{2.7}$$

Based on the above relationships, the wettability of machined surfaces can be quantitatively compared. Another important criterion for evaluating SMT is anisotropic wettability, denoted as ($\Delta\theta$). Anisotropic wettability refers to the variation in contact angle measurements across different directions on the surface. It is inversely related to overall wettability: the smaller the $\Delta\theta$, the greater the surface wettability [24]. According to Eq. 2.8, anisotropic wettability is calculated as the difference in contact angles measured in the feed direction (θ_f) and the cutting speed direction (θ_c).

$$\Delta\theta = \theta_f - \theta_c \tag{2.8}$$

Lotfi et al. [25] carried out face turning operations using both conventional machining and three-dimensional ultrasonic vibrations. The machining parameters were as follows: a depth of cut of 0.5 mm, a feed rate of 0.11 mm/rev, and a cutting speed of 60 m/min. Under ultrasonic conditions, the vibration amplitude (peak-to-peak) was approximately 12 μm in all directions. The turning operations were conducted in dry conditions, and the workpiece material used was Grade-5 titanium alloy. Following the machining process, surface topographies were analyzed using an optical microscope (for 2D imaging) and a nano-focus microscope (for 3D imaging). As observed in Fig. 2.7, the surface topography changed significantly when three-dimensional ultrasonic vibrations were applied. In the conventional case, surface roughness followed the direction of tool feed. In contrast, UAM not only altered the roughness pattern but also generated semi-spherical surface textures. Moreover, surface roughness was notably reduced, from 1.8 μm in the conventional case to 1.1 μm under ultrasonic conditions. This improvement is attributed to the additional vibratory motion of the cutting tool, which effectively flattened the surface peaks generated during conventional machining. Another important observation from the results is the improved surface integration, which was further evaluated using wettability testing. To assess this, a CA-ES10 contact angle measurement device was used, and the results are presented in Fig. 2.8. From these images, water droplets were placed on the machined surfaces to measure the contact angles in both the feed (θ_f) and cutting speed (θ_c) directions. For both conventional machining and UAM, the contact angle in the feed direction was greater than in the cutting speed direction. Notably, these contact angles were lower in the ultrasonic vibration mode compared to conventional machining. This indicates that ultrasonic vibrations effectively mitigate the directional influence of the tool feed on the surface wettability to a significant degree. Furthermore, the results show that the anisotropic wettability in the vibratory mode ($\Delta\theta = 5.4°$) is significantly lower than that in conventional machining ($\Delta\theta = 10.7°$) when comparing the contact angle differences between the feed and cutting speed directions. As previously noted, a lower anisotropic wettability corresponds to higher overall wettability. These findings collectively demonstrate that the application of three-dimensional ultrasonic vibrations enhances the SMT and improves the uniformity and integrity of the machined surfaces.

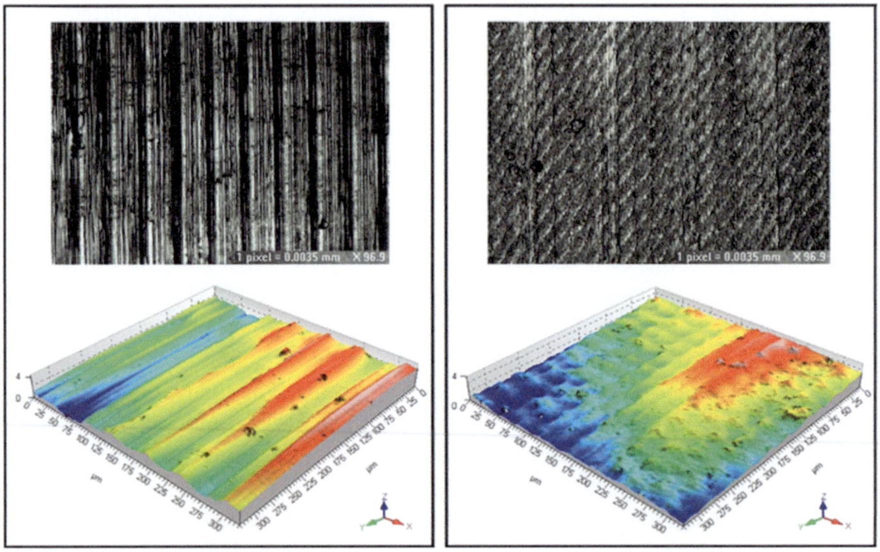

(a) (b)

Fig. 2.7 2D and 3D images of SMT obtained for **a** conventional machining and **b** UAM [25]

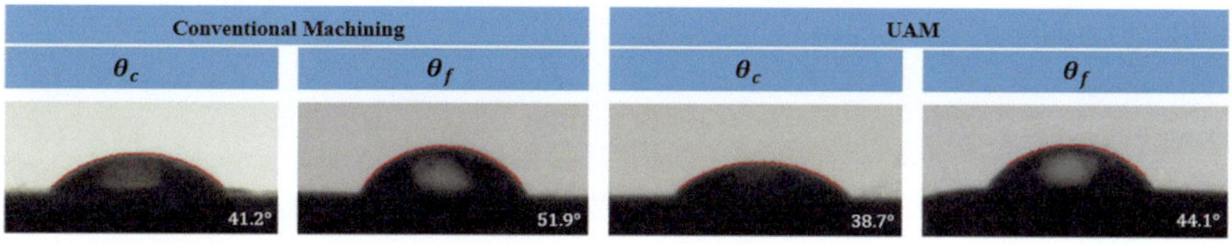

Fig. 2.8 Wettability results for conventional machining and UAM [25]

2.4 Surface Micro-topography Applications

2.4.1 Industrial Applications

The application of microtextured surfaces in industry has advanced significantly more than in the medical field. These applications include reducing grain size, enhancing microhardness, decreasing wear and friction, and improving lubrication performance in industrial components.

Microstructure and microhardness, along with SMT, are key indicators of surface integrity and are interrelated [26]. Typically, the microstructure of a material is evaluated by comparing grain size before and after machining. The primary parameters influencing microstructure include: (i) temperature, (ii) strain, and (iii) strain rate [27]. Variations in any of these factors can significantly alter grain size. According to the Hall–Petch relationship given in Eq. 2.9, a decrease in grain size leads to an increase in material hardness. This relationship involves the yield stress (σ_o), frictional stress (σ_i), grain size (d), and material constant (k). Consequently, modifications in SMT can influence grain size and, in turn, affect the microhardness of the material surface [28].

$$\sigma_o = \sigma_i + \frac{k}{\sqrt{d}} \tag{2.9}$$

In a recent study [29], surface integrity and microstructural changes in ultrasonic-assisted turning of titanium alloys were investigated. Grain size and microhardness were measured during conventional and ultrasonic-assisted turning operations. The experimental procedures were conducted according to the conditions summarized in Table 2.1. The peak-to-peak vibration amplitudes were 14 μm in the depth-of-cut direction, 12 μm in the feed direction, and 14 μm in the cutting speed direction. To evaluate the grain size and surface microhardness, a scanning electron microscope (SEM) and a microhardness tester were employed, respectively. Moreover, a metallographic preparation process was performed prior to SEM imaging. Figure 2.9 shows the microstructural views and macro-images of the machined surfaces.

Figure 2.10 shows the average grain size and microhardness results for conventional and ultrasonic-assisted turning processes. Considering that the initial grain size

Table 2.1 Design of experiments for ultrasonic-assisted turning tests

Test number	1	2	3	4	5	6	7	8	9	10	11	12
Cutting speed (m/min)	15	30	60	15	30	60	15	30	60	15	30	60
Feed rate (mm/rev)	0.05	0.05	0.05	0.11	0.11	0.11	0.05	0.05	0.05	0.11	0.11	0.11
Depth of cut (mm)	0.5	0.5	0.5	0.5	0.5	0.5	1	1	1	1	1	1

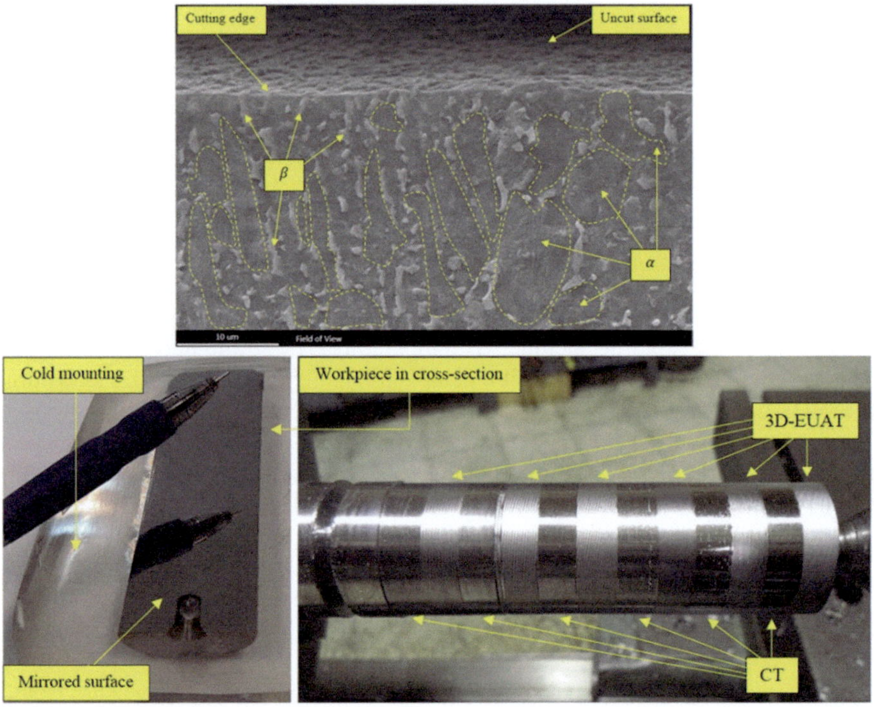

Fig. 2.9 Microstructure and macro-images of the machined surfaces [29]

of the titanium alloy before machining is 15 μm, both conventional and ultrasonic-assisted turning operations result in grain size reduction. Across all experiments, ultrasonic vibrations lead to a greater reduction in surface grain size compared to conventional turning. Specifically, the grain size in the ultrasonic mode is approximately 5% smaller than in the conventional mode. Among the machining parameters, cutting speed has the most significant influence on grain size reduction. It is noteworthy that three-dimensional tool vibrations, through micron-level penetration into the workpiece, generate microtextures on the surface and contribute to refining the titanium microstructure. These changes also influence surface microhardness.

According to the microhardness results, the initial microhardness of titanium is 360 HV, an increase in microhardness is observed in both conventional and ultrasonic-assisted turning operations. The results indicate that ultrasonic vibrations lead to a greater increase in microhardness than conventional mode. This improvement may be attributed to the finer surface microstructure produced by superimposing ultrasonic vibrations onto the cutting tool. Consequently, creating SMT with the help of ultrasonic vibrations not only changes the surface characteristics of the workpiece but also causes microstructural changes beneath the surface, including grain size reduction and increased microhardness. Therefore, this method can be effectively employed to enhance the mechanical properties of machined parts in industrial applications.

Wear and friction are closely related parameters and are best examined together. Friction refers to the resistive force that occurs at the interface of two objects in relative motion [30]. Wear, on the other hand, is a phenomenon that impacts the quality and longevity of surfaces. It refers to the surface damage that results from mechanical or chemical interactions during operation [31]. By increasing the wear resistance of industrial components, the service life of these parts can be significantly extended.

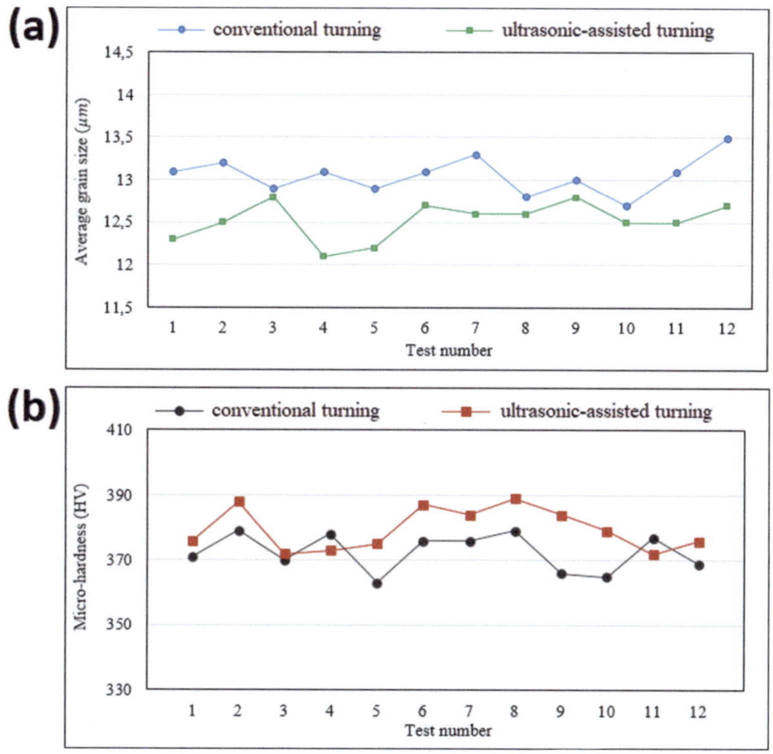

Fig. 2.10 **a** Average grain size and **b** microhardness for conventional and ultrasonic-assisted turning [29]

Table 2.2 Process parameters for turning operations

Test number	1	2	3	4	5	6	7	8	9
Cutting speed (m/min)	15	30	60	15	30	60	15	30	60
Feed rate (mm/rev)	0.05	0.05	0.05	0.11	0.11	0.11	0.05	0.05	0.05

Amini et al. [32] investigated the effects of microtextured surfaces generated by UAM on the friction and wear performance of machined components. In this experimental study, a face turning process was conducted, involving four types of tests: conventional turning (CT), linear vibratory turning (LVT), elliptical vibratory turning (EVT), and three-dimensional vibratory turning (3D-VT). The tests were performed on AA7075-T6 workpiece material. The cutting parameters are presented in Table 2.2, with a constant depth of cut of 0.25 mm for all cutting conditions. After completing the face turning operations, a pin-on-disk apparatus was used to analyze both friction and wear. During the tribological tests, a load of 10 N was applied, the pin had a diameter of 6 mm, and the sliding distance was set to 300 mm.

From the results, the application of ultrasonic vibrations results in machined surfaces exhibiting significantly lower friction coefficients compared to conventional turning. Among the different vibration modes, the lowest friction is observed when three-dimensional vibrations are employed. Furthermore, increasing the cutting speed and feed rate leads to a further reduction in friction coefficients. According to Fig. 2.11, which shows the wear test results, surfaces machined with three-dimensional vibrations experience the least wear. This trend continues with elliptical vibratory turning, followed by linear vibratory turning, and lastly, conventional turning, which shows the highest wear. These findings indicate that ultrasonic vibrations play a critical role in minimizing both friction and wear on machined surfaces. The underlying reason for this improvement lies in the formation of SMT. Specifically, ultrasonic vibrations generate microtextures that reduce the real area of contact between surfaces. This reduction in contact area lowers the frictional forces during sliding interactions and decreases the extent of surface wear. Consequently, the use of ultrasonic-assisted turning not only improves the surface characteristics but also enhances the functional performance and longevity of machined components.

2.4.2 Medical Applications

One of the most significant applications of SMT in the medical field is in the development of medical implants. Although modern implants offer acceptable levels of durability and promote cell growth, ongoing studies continue to seek enhancements in their performance and longevity [33]. A substantial part of these studies is centered on improving osseointegration, which is the direct structural and functional connection between living bone and the surface of a load-bearing implant. Key factors influencing osseointegration include the enhancement of the hydrophilicity of the

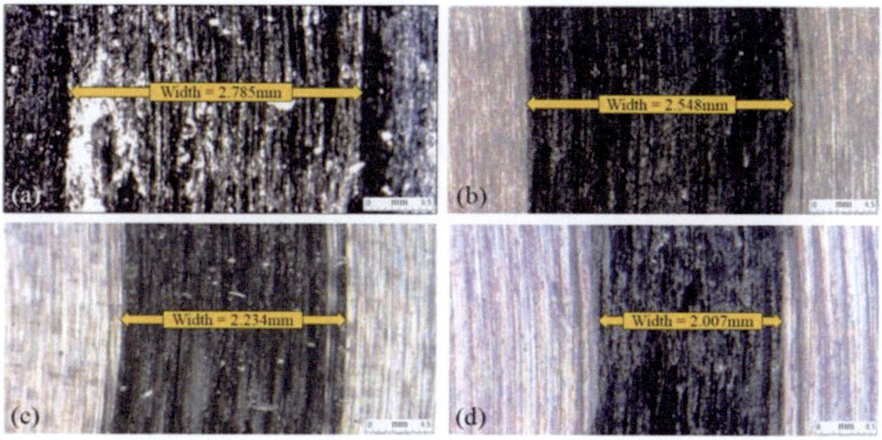

Fig. 2.11 Wear channels on the machined surfaces after **a** CT, **b** LVT, **c** EVT, and **d** 3D-VT [32]

implant surface and the increase in the contact area between the implant and the surrounding bone tissue [34]. These aspects are directly influenced by the surface characteristics of the implant, which are in turn affected by SMT. Therefore, osseointegration cannot be investigated in isolation; it is closely linked to the wettability of the implant surface, the type of material used, and the rate of bone cell growth. Together, these parameters determine the quality and overall effectiveness of medical implants. As such, an in-depth understanding of the implant material and the biological response—particularly bone cell proliferation—is essential for advancing the design and functionality of next-generation implants.

Titanium Grade-5 is the most commonly used material in the manufacturing of medical implants due to its high corrosion resistance and favorable strength-to-weight ratio, making it an ideal substitute for natural bone tissue [35]. Despite these advantages, titanium surfaces face a notable limitation: biological inertness, which results in inadequate bonding between the implant and surrounding biological tissues [36]. This challenge can be addressed through surface modification, particularly by altering the SMT. The physicochemical properties of titanium, such as SMT, surface energy, electrical charge, and wettability, are closely interlinked and have a direct impact on key biological responses, including cell adhesion, orientation, motility, and proliferation. These cellular interactions occur at the interface between the implant and the host tissue, highlighting the critical role of surface design [37]. Microtexturing the surface enhances these interactions by modifying the surface geometry, thereby influencing hydrophilicity and improving overall biological performance. Although titanium is inherently hydrophilic, increasing its effective surface area through microtexturing further enhances this property, thereby promoting more favorable biological outcomes [38].

Bone tissue is a composite biological material composed primarily of minerals, proteins, and water, playing critical roles in the human body such as providing structural support, serving as a reservoir for minerals, and producing blood platelets

[39]. This tissue consists of four primary cell types: progenitor cells, osteoblasts, osteocytes, and osteoclasts. Progenitor cells differentiate into osteoblasts, which are responsible for the synthesis of new bone tissue and the production of the extracellular matrix. Osteocytes are mature bone cells involved in the maintenance of bone, while osteoclasts facilitate bone resorption. While all four cell types function to maintain bone homeostasis, the activity of osteoblasts notably increases in scenarios such as fracture healing or implant integration, as they are specialized in bone formation and cell proliferation, with the term cell growth referring to both an increase in cell size and number [40]. In addition to its cellular components, bone tissue is largely composed of an extracellular matrix, which is filled with large molecules that provide structural and biochemical support. These molecules are classified into polysaccharide chains and fibrous proteins, the latter being crucial for cell adhesion within the matrix [41]. Given this intricate biological structure, recent studies have focused on modifying the surface topography of titanium implants to improve osteoblast cell adhesion and proliferation. By introducing microtextures onto the titanium surface, the hydrophilicity of the material is significantly enhanced, creating an environment conducive to osteoblast growth and activity. This enhancement supports more effective and accelerated osseointegration, ultimately improving the performance and longevity of medical implants.

In a recent study [42], an experimental investigation was conducted using a rotary UAM tool capable of generating longitudinal vibrations during both drilling and milling operations, as shown in Fig. 2.12. The objective was to evaluate the biological characteristics of surfaces machined under conventional milling and ultrasonic-assisted milling conditions. The cutting experiments were designed based on the parameters listed in Table 2.3, involving a total of 12 test conditions. The workpiece material used was titanium alloy (Grade-5), which is commonly employed in medical applications due to its biocompatibility. A 10 mm diameter milling tool was used, with a constant depth of cut of 0.5 mm and a vibration amplitude of 0.6 μm in the ultrasonic-assisted mode. To assess the surface quality and biological response, SEM was employed for detailed imaging. For the cell culture tests, the MC3T3 cell line, derived from mouse calvarial bone tissue, was selected. These cells are known as osteoblast progenitor cells, making them highly relevant for evaluating the osseointegration potential and biocompatibility of the machined surfaces.

As shown in Fig. 2.13, microtextures are generated on the titanium surface during ultrasonic-assisted milling, whereas the surface produced by conventional milling retains typical linear peaks and valleys. Cell culture experiments conducted on these machined surfaces revealed distinct differences in osteoblast behavior. Specifically, the images show that in the conventional milling condition, cells tend to grow in clusters, indicating stronger cell-to-cell communication. In contrast, under the ultrasonic-assisted milling condition, cells predominantly appear singly dispersed, suggesting that they encountered less resistance in adhering to the microtextured titanium surface. This observation implies that the presence of SMT, produced via ultrasonic-assisted milling, facilitates initial cell attachment and creates a more biocompatible environment for bone cell growth. Therefore, microtexturing through ultrasonic vibrations enhances surface wettability and significantly improves the

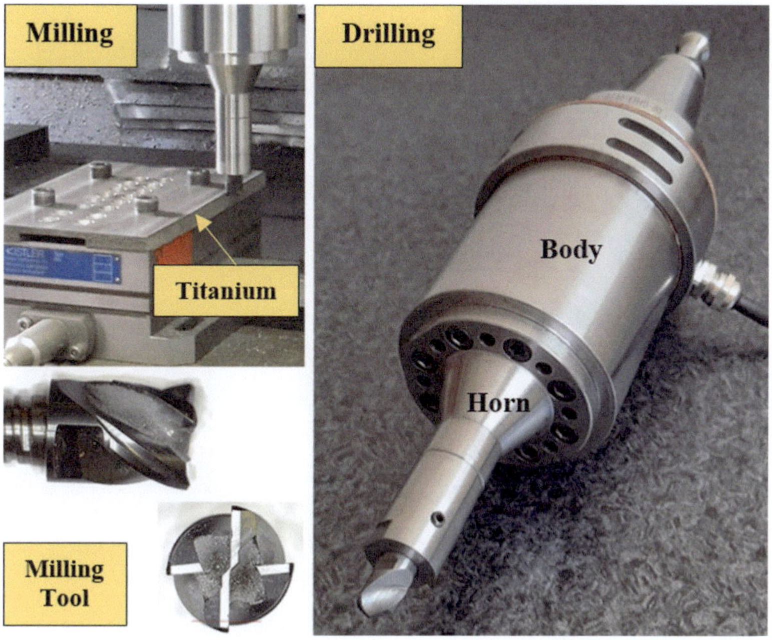

Fig. 2.12 A rotary UAM tool capable of generating longitudinal vibration in drilling and milling processes [42]

Table 2.3 Machining parameters in rotary UAM

Test number	1	2	3	4	5	6
Rotational speed (rpm)	1000	1500	2000	1000	1500	2000
Feed speed (mm/min)	5	5	5	10	10	10

potential for osseointegration, making it a promising approach for the development of high-performance medical implants [43, 44].

2.5 Conclusions

In this chapter, the phenomenon of SMT in UAM processes was comprehensively investigated. The discussion began with an overview of the mechanisms of ultrasonic vibrations—including linear, elliptical, and three-dimensional modes—as applied in machining operations such as turning, drilling, and milling. This was followed by an in-depth exploration of SMT-related parameters, including contact angle, wettability, and anisotropic wettability. Subsequently, the industrial and biomedical applications

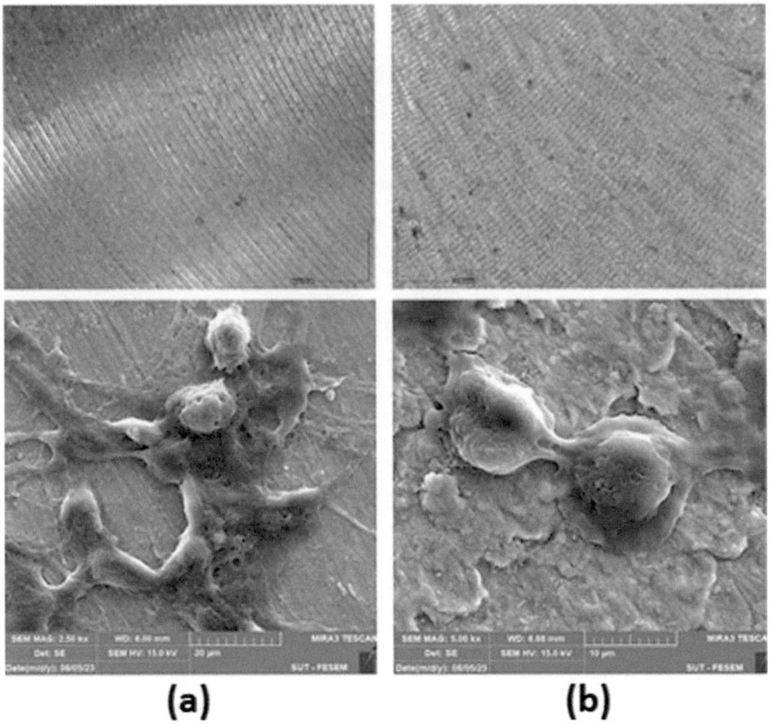

Fig. 2.13 Machined surfaces and cell growth on the surfaces after **a** conventional milling and **b** ultrasonic-assisted milling [42]

of SMT were reviewed, emphasizing its functional advantages. The key findings are summarized as follows:

- The analysis of contact angle, wettability, and anisotropic wettability demonstrated that three-dimensional ultrasonic vibrations significantly enhance SMT and improve the surface integrity of machined components.
- The application of UAM results in both surface and subsurface modifications, particularly a reduction in grain size, which subsequently leads to an increase in microhardness, and thereby improving the mechanical performance of the surface.
- The introduction of microtextures through ultrasonic vibrations was shown to reduce the effective contact area during mechanical interaction, leading to a notable reduction in friction and wear.
- In biomedical applications, specifically in the fabrication of titanium implants, the creation of microtextures enhances surface hydrophilicity, which facilitates osteoblast cell attachment and growth. This improvement in biological compatibility supports more effective osseointegration, a critical factor in implant success.

Overall, the use of ultrasonic vibrations in machining processes presents a promising strategy for SMT that improves both mechanical properties and biological responses, thereby broadening the scope of UAM in micromanufacturing and medical applications.

References

1. Tang W, Zhou Y, Zhu H, Yang H. The effect of surface texturing on reducing the friction and wear of steel under lubricated sliding contact. Applied surface science. 2013 May 15;273:199–204.
2. Sajjady SA, Abadi HN, Amini S, Nosouhi R. Analytical and experimental study of topography of surface texture in ultrasonic vibration assisted turning. Materials & Design. 2016 Mar 5;93:311–23.
3. Dhananchezian M. Effectiveness of cryogenic cooling in turning of Inconel 625 alloy. In Advances in Manufacturing Processes: Select Proceedings of ICEMMM 2018 2019 (pp. 591–597). Springer Singapore.
4. Pradhan S, Singh S, Prakash C, Królczyk G, Pramanik A, Pruncu CI. Investigation of machining characteristics of hard-to-machine Ti-6Al-4V-ELI alloy for biomedical applications. Journal of Materials Research and Technology. 2019 Sep 1;8(5):4849–62.
5. Lotfi M, Farid AA, Akbari J, Sharif S, Mohruni AS. Evaluation of surface integrity when drilling Inconel 718 through experimental measurement and finite element analysis. The International Journal of Advanced Manufacturing Technology. 2022 Apr 1:1–2.
6. Lotfi M, Amini S. FE simulation of linear and elliptical ultrasonic vibrations in turning of Inconel 718. Proceedings of the Institution of Mechanical Engineers, Part E: Journal of Process Mechanical Engineering. 2018 Aug;232(4):438–48.
7. Lotfi M, Amini S. Effect of ultrasonic vibration on frictional behavior of tool–chip interface: Finite element analysis and experimental study. Proceedings of the Institution of Mechanical Engineers, Part B: Journal of Engineering Manufacture. 2018 May;232(7):1212–20.
8. Amini S, Alinaghian I, Lotfi M, Teimouri R, Alinaghian M. Modified drilling process of AISI 1045 steel: a hybrid optimization. Engineering Science and Technology, an International Journal. 2017 Dec 1;20(6):1653–61.
9. Amini S, Aghaei M, Lotfi M, Hakimi E. Analysis of linear vibration in rotary turning of AISI 4140 steel. The International Journal of Advanced Manufacturing Technology. 2017 Aug;91:4107–16.
10. Lotfi M, Amini S, Ashrafi H. Theoretical and numerical modeling of tool–chip friction in ultrasonic-assisted turning. Proceedings of the Institution of Mechanical Engineers, Part E: Journal of Process Mechanical Engineering. 2019 Aug;233(4):824–38.
11. Lotfi M, Amini S, Sajjady SA. Development of a friction model based on oblique cutting theory. International Journal of Mechanical Sciences. 2019 Sep 1;160:241–54.
12. Lotfi M. Effect of 3D Elliptical Vibrations on Tool-chip Friction and Microstructure Changes in the Machining of Titanium Alloy [dissertation]. Iran: University of Kashan; 2019.
13. Amini S, Soleimani M, Paktinat H, Lotfi M. Effect of longitudinal– torsional vibration in ultrasonic-assisted drilling. Materials and Manufacturing Processes. 2017 Apr 26;32(6):616–22.
14. Amirnia M, Akbari J, Lotfi M. Ultrasonic lubrication assisted milling of CFRP: Elimination of fiber pull-out. Proceedings of the Institution of Mechanical Engineers, Part E: Journal of Process Mechanical Engineering. 2022 Dec;236(6):2651–7.
15. Lotfi M, Akbari J. Finite element simulation of ultrasonic-assisted machining: a review. The International Journal of Advanced Manufacturing Technology. 2021 Oct;116:2777–96.

16. Ali S, Kurniawan R, Ko TJ. Development of 3D resonant elliptical vibration transducer for dual-frequency micro-dimple surface texturing. International Journal of Precision Engineering and Manufacturing. 2021 Aug;22(8):1365–79.
17. Dubey V, Sharma AK, Pimenov DY. Prediction of surface roughness using machine learning approach in MQL turning of AISI 304 steel by varying nanoparticle size in the cutting fluid. Lubricants. 2022 May 2;10(5):81.
18. Lotfi M, Amini S, Aghaei M. 3D analysis of surface topography in vibratory turning. The International Journal of Advanced Manufacturing Technology. 2018 Mar;95:197–204.
19. Sajjady SA, Lotfi M, Amini S, Toutounchi H, Bami AB. Improving the surface energy of titanium implants by the creation of hierarchical textures on the surface via three-dimensional elliptical vibration turning for enhanced osseointegration. Proceedings of the Institution of Mechanical Engineers, Part H: Journal of Engineering in Medicine. 2019 Dec;233(12):1226–36.
20. Lotfi M, Amini S, Aghayar Z, Sajjady SA, Farid AA. Effect of 3D elliptical ultrasonic assisted boring on surface integrity. Measurement. 2020 Oct 15;163:108008.
21. Lotfi M, Ashrafi H, Amini S, Akhavan Farid A, Jahanbakhsh M. Characterization of various coatings on wear suppression in turning of Inconel 625: A three-dimensional numerical simulation. Proceedings of the Institution of Mechanical Engineers, Part J: Journal of Engineering Tribology. 2017 Jun;231(6):734–44.
22. Hosseinabadi HN, Sajjady SA, Amini S. Creating micro textured surfaces for the improvement of surface wettability through ultrasonic vibration assisted turning. The International Journal of Advanced Manufacturing Technology. 2018 May;96:2825–39.
23. Yuan Y, Lee TR. Contact angle and wetting properties. In Surface science techniques 2013 Jan 1 (pp. 3–34). Berlin, Heidelberg: Springer Berlin Heidelberg.
24. Kurniawan R, Ali S, Ko TJ. Measurement of wettability on rhombohedral pattern fabricated by using 3D-UEVT. Measurement. 2020 Aug 1;160:107784.
25. Lotfi M, Sajjady SA, Amini S. Wettability analysis of titanium alloy in 3D elliptical ultrasonic assisted turning. International Journal of Lightweight Materials and Manufacture. 2019 Sep 1;2(3):235–40.
26. Sun Z, Liu Y, Geng D, Zhang D, Ying E, Liu R, Jiang X. Cutting performance and surface integrity during rotary ultrasonic elliptical milling of cast Ni-based superalloy. Journal of Materials Research and Technology. 2025 Jan 11.
27. Wang X, Zhuang B, Wu T, Su H, Chen Y, Zhao B, Ding W. Microstructure evolution characterization of Ti2AlNb intermetallic alloys subjected to side milling. Materials Today Communications. 2025 Jan 1;42:111257.
28. Jafarian F, Ciaran MI, Umbrello D, Arrazola PJ, Filice L, Amirabadi H. Finite element simulation of machining Inconel 718 alloy including microstructure changes. International Journal of Mechanical Sciences. 2014 Nov 1;88:110–21.
29. Lotfi M, Amini S, Akbari J. Surface integrity and microstructure changes in 3D elliptical ultrasonic assisted turning of Ti–6Al–4V: FEM and experimental examination. Tribology International. 2020 Nov 1;151:106492.
30. Lu Y, Han D, Fu Q, Lu X, Zhang Y, Wei Z, Chen Y. Experimental investigation of stick-slip behaviors in dry sliding friction. Tribology International. 2025 Jan 1;201:110221.
31. Amini S, Abbaszadeh S, Lotfi M. Measuring wear-resistance of AISI 1.7225 steel under various heat treatments: Hydraulic spool valve. Measurement. 2017 Feb 1;98:179–85.
32. Amini S, Hosseinabadi HN, Sajjady SA. Experimental study on effect of micro textured surfaces generated by ultrasonic vibration assisted face turning on friction and wear performance. Applied Surface Science. 2016 Dec 30;390:633–48.
33. Pandey A, Sahoo S. Progress on medical implant: a review and prospects. Journal of Bionic Engineering. 2023 Mar;20(2):470–94.
34. Jacobs TW, Dillon JT, Cohen DJ, Boyan BD, Schwartz Z. Different Methods to Modify the Hydrophilicity of Titanium Implants with Biomimetic Surface Topography to Induce Variable Responses in Bone Marrow Stromal Cells. Biomimetics. 2024 Apr 10;9(4):227.

35. Najafizadeh M, Yazdi S, Bozorg M, Ghasempour-Mouziraji M, Hosseinzadeh M, Zarrabian M, Cavaliere P. Classification and applications of titanium and its alloys: A review. Journal of Alloys and Compounds Communications. 2024 Aug 2:100019.
36. Khang D, Lu J, Yao C, Haberstroh KM, Webster TJ. The role of nanometer and sub-micron surface features on vascular and bone cell adhesion on titanium. Biomaterials. 2008 Mar 1;29(8):970–83.
37. Li H, Cong Y, Zhou S, Zhang J. Cutting-Based Manufacturing and Surface Wettability of Microtextures on Pure Titanium. Materials. 2024 Aug 4;17(15):3861.
38. Szymańska J, Krzywicka M, Kobus Z, Malm A, Grzegorczyk A. The Influence of Selected Titanium Alloy Micro-Texture Parameters on Bacterial Adhesion. Materials. 2024 Sep 28;17(19):4765.
39. Florencio-Silva R, Sasso GR, Sasso-Cerri E, Simões MJ, Cerri PS. Biology of bone tissue: structure, function, and factors that influence bone cells. BioMed research international. 2015;2015(1):421746.
40. Shestopaloff YK. Why cells grow and divide? General growth mechanism and how it defines cells' growth, reproduction and metabolic properties. Biophysical Reviews and Letters. 2015 Dec 22;10(04):209–56.
41. Perić Kačarević Ž, Rider P, Alkildani S, Retnasingh S, Pejakić M, Schnettler R, Gosau M, Smeets R, Jung O, Barbeck M. An introduction to bone tissue engineering. The International Journal of Artificial Organs. 2020 Feb;43(2):69–86.
42. Lotfi M, Charkhian A, Akbari J. Surface analysis in rotary ultrasonic-assisted milling of CFRP and titanium. Journal of Manufacturing Processes. 2022 Dec 1;84:174–82.
43. Ruhibakhsh R. Implementation of vibration assisted milling on titanium and investigation of its effects on surface texture and biocompatibility [master's thesis]. Iran: Sharif University of Technology; 2024.
44. Zamani M, Farahnakian M, Elhami S. Employment of ultrasonic assisted turning in the fabrication of microtextures to improve the surface adhesion of the titanium implant. Proceedings of the Institution of Mechanical Engineers, Part B: Journal of Engineering Manufacture. 2021 Oct;235(12):1983–91.

Chapter 3
Microtexturing for Performance Enhancement of Carbide Cutting Tools

Juan Manuel Vazquez-Martinez, Moises Batista, Javier Garcia-Fernandez, Irene del Sol, and Jorge Salguero

Abstract Microtexturing has emerged as a highly effective technique for enhancing the performance of carbide cutting tools by precisely altering their surface characteristics at the microscale. Strategically designed microtextures can significantly improve the tribological behavior of cutting tools by reducing friction, facilitating efficient chip evacuation, and minimizing tool wear during machining operations. Practical implementations have demonstrated notable improvements in tool life, machining efficiency, and the surface finish of workpieces. This chapter presents an overview of microtexturing on carbide cutting tools to enhance machining performance. It discusses the fundamental principles, methodologies, and performance outcomes associated with surface texturing techniques applied to carbide cutting tools.

Keywords Microtexture design · Cutting tool performance · Friction and wear control · Surface engineering · Material removal

3.1 Introduction

The performance of carbide cutting tools in machining operations is strongly influenced by the nature of the workpiece material, cutting conditions, and the selection of operational parameters. These tools are engineered to endure harsh environments characterized by elevated temperatures, high mechanical stresses, and severe wear—common in the machining of hard, tough, or abrasive materials such as alloy steels, superalloys, and metallic composites. Process variables such as cutting speed, feed rate, depth of cut, and the application of coolants directly affect heat generation,

J. M. Vazquez-Martinez (✉) · M. Batista · J. Garcia-Fernandez · I. del Sol · J. Salguero
Department of Mechanical Engineering and Industrial Design, School of Engineering, University of Cadiz, Puerto Real (Cadiz), Spain
e-mail: juanmanuel.vazquez@uca.es

S. Gürgen (ed.), *Micromanufacturing Operations*,
https://doi.org/10.1007/978-3-032-04193-7_3

plastic deformation, and wear mechanisms, including cracking, abrasion, and adhesion. Optimal selection of cutting tools and machining parameters is crucial for maximizing tool life, maintaining dimensional accuracy, and ensuring high performance under demanding machining conditions.

Carbide cutting tools—typically composed of tungsten carbide reinforced with a cobalt binder—play a critical role in modern manufacturing due to their exceptional hardness, wear resistance, and thermal stability. These tools are specifically engineered to withstand the intense mechanical and thermal stresses encountered during high-speed machining, cutting, and shaping operations. They are widely used across industries such as automotive, aerospace, construction, and mold-making. The ability of tungsten carbide tools to retain dimensional precision and operational reliability under such demanding conditions makes them indispensable for achieving high-quality and consistent production outcomes [1, 2].

In many machining processes, the primary function of cutting tools is to efficiently remove material from workpieces while maintaining tight dimensional tolerances and high-quality surface finishes. Achieving this requires tools capable of resisting deformation and wear under elevated temperatures and substantial mechanical loads. Tungsten carbide tools meet these demands, often outperforming alternative materials such as high-speed steel—especially in challenging applications like milling, turning, and drilling of hard or abrasive materials. However, the performance of tungsten carbide tools is still affected by factors such as tool geometry, material composition, and surface condition. Common wear mechanisms—including abrasion, adhesion, oxidation, and thermal fatigue—can substantially shorten tool life, leading to higher operational costs and increased downtime. Furthermore, the growing emphasis on productivity, energy efficiency, and sustainability has intensified the need for innovative strategies to enhance the performance and longevity of cutting tools [2]. In this context, surface texturing has emerged as a promising strategy for enhancing the performance of tungsten carbide tools. By introducing microscale or nanoscale patterns onto the tool surface, surface texturing can effectively reduce friction, improve wear resistance, and facilitate better chip evacuation. These advancements directly address the challenges posed by modern machining demands, providing a viable solution to further optimize the efficiency and longevity of tungsten carbide cutting tools in industrial applications.

Optimizing cutting tool performance is a critical challenge in modern manufacturing, driven by the need to balance productivity, precision, and cost-effectiveness. Cutting tools, particularly those made from tungsten carbide, endure extreme mechanical and thermal stresses during machining operations. These harsh conditions can result in rapid wear, tool failure, and inconsistent machining outcomes, all of which directly impact production efficiency and product quality. To address these challenges, it is essential to develop a comprehensive understanding of the wear mechanisms and factors that influence tool performance. As shown in Fig. 3.1, one of the primary challenges in machining is wear resistance, as cutting tools are continuously subjected to frictional interactions with the workpieces. Abrasive wear occurs when the tool interacts with hard workpiece materials, leading to material loss at the cutting edge. Adhesive wear, in contrast, results from material transfer between the

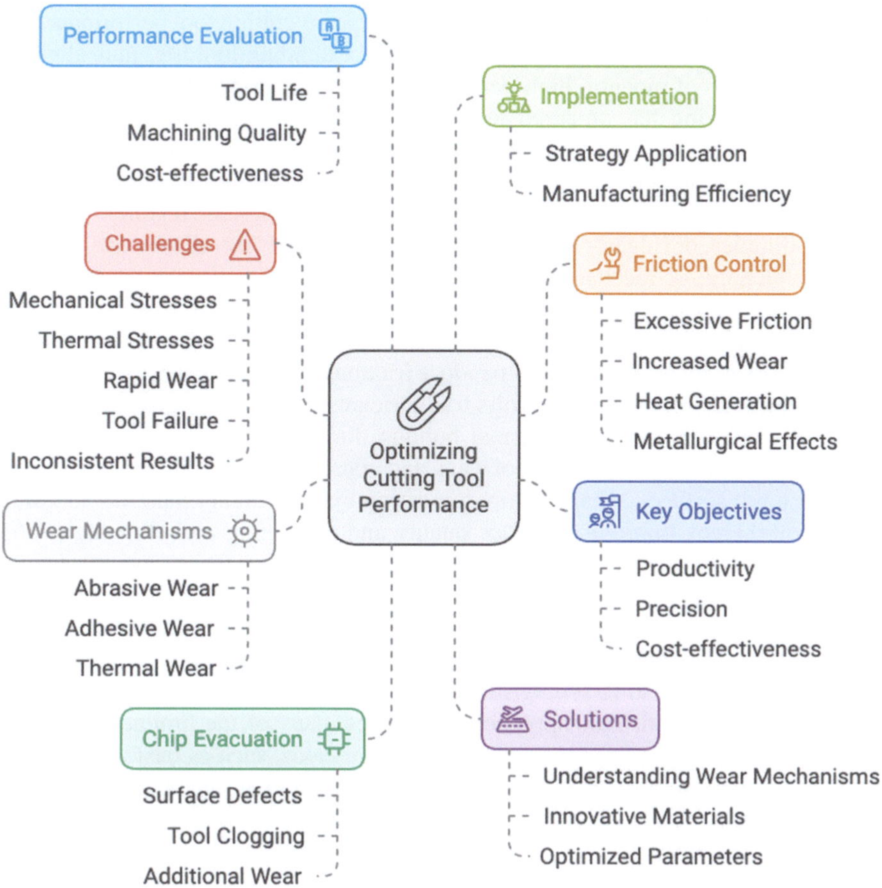

Fig. 3.1 Challenges in the optimization of cutting tool performance

tool and the workpiece, causing surface damage and deterioration. Furthermore, high-temperature machining conditions introduce oxidation and thermal fatigue, which accelerate tool degradation and shorten its effective lifespan. Another major challenge is controlling friction at the tool–workpiece interface. Excessive friction not only accelerates tool wear but also generates heat, which can adversely affect the metallurgical properties of both the tool and the workpiece. Moreover, efficient chip evacuation is critical in high-speed machining, as poor chip removal can lead to surface defects, tool clogging, and further wear [3, 4]. The pursuit of optimization is further complicated by evolving industry demands, including the machining of harder and more complex materials, achieving tighter tolerances, and minimizing environmental impact. While traditional methods, such as adjusting tool geometry or applying coatings, have yielded incremental improvements, they are often inadequate to meet the increasingly stringent performance requirements. To address these challenges, advanced techniques such as surface texturing have attracted significant attention. Surface texturing has the potential to increase cutting tool performance by enhancing wear resistance, reducing friction, and extending tool life, thus aligning with the demands of modern, high-performance manufacturing environments.

Surface texturing has emerged as a good solution for improving the performance of cutting tools, driven by the increasing demands for efficiency, precision, and durability in modern manufacturing. It involves the deliberate modification of tool surfaces through the introduction of microscale or nanoscale features—such as grooves, dimples, or ridges—using advanced methods like laser machining, electrochemical etching, or mechanical micromachining. These engineered textures are designed to interact with the machining environment, enhancing the functional performance of tools under demanding operating conditions. The primary advantage of surface texturing lies in its ability to reduce friction at the tool–workpiece interface. Micropatterns serve as microreservoirs for lubricants, enabling localized lubrication that lowers contact stress and thermal buildup during machining. This results in reduced cutting forces, decreased tool wear, and improved energy efficiency. Furthermore, textured surfaces enhance chip evacuation by minimizing material adhesion to the tool, thereby improving surface quality and preventing tool clogging [5, 6]. Surface texturing also enhances wear resistance by modifying the contact mechanics between the tool and the workpiece. Certain texture patterns can inhibit crack propagation and reduce the buildup of thermal stresses—both critical contributors to tool degradation. When combined with advanced coatings, surface texturing can further extend tool life by creating synergistic effects that enhance resistance to oxidation, abrasion, and adhesion. This approach overcomes many of the limitations associated with traditional performance enhancement methods, such as basic coatings or geometric modifications. By enabling precise control over surface microstructures, surface texturing offers a versatile and customizable solution tailored to a wide range of machining applications.

3.2 An Overview of Tungsten Carbide

Tungsten carbide, also referred to as cemented carbide, is renowned for its exceptional hardness, approaching that of diamond, along with its excellent wear resistance and thermal stability. These properties make it indispensable in a wide range of industrial applications, particularly in high-performance machining. As shown in Fig. 3.2, the material consists of hard tungsten carbide particles bonded together by a metallic binder, typically cobalt [7]. Tungsten carbide is produced through powder metallurgy, a process that minimizes material waste and enhances production efficiency. However, this method also raises environmental and occupational health concerns, particularly due to the handling of cobalt and other fine metal powders. Cobalt is linked to serious respiratory conditions in exposed workers, including interstitial pulmonary fibrosis and giant cell interstitial pneumonia [8]. As a result, stringent safety regulations have been implemented to mitigate exposure risks during the manufacturing and handling of tungsten carbide in industrial environments [9].

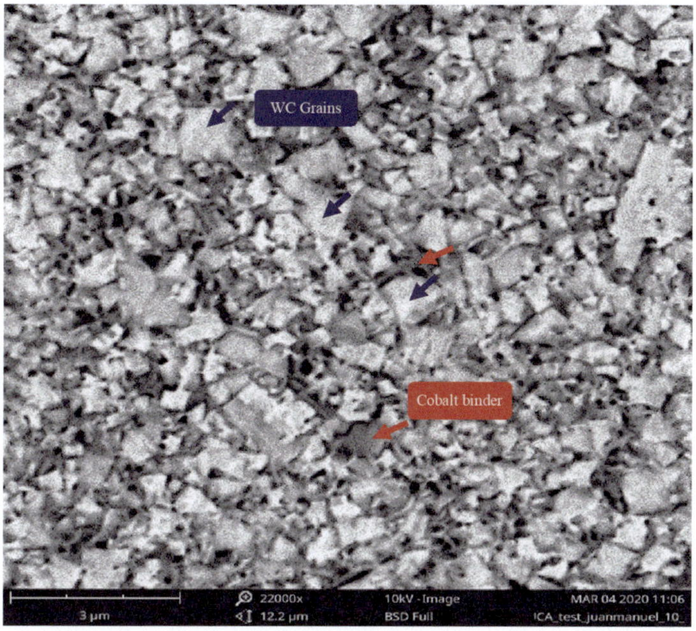

Fig. 3.2 Microstructure of a tungsten carbide tool

Tungsten carbide is also a high-cost material, primarily due to the expense and scarcity of its constituent elements. Tungsten, known for having one of the highest melting points among metals, is challenging and energy-intensive to process. Cobalt, used as a binder, is not only costly but also subject to supply volatility, as its availability is heavily influenced by geopolitical factors. Moreover, the powder metallurgy process required to produce tungsten carbide, while essential for achieving its superior mechanical properties, involves complex and resource-intensive techniques that further contribute to its cost [8]. Given these economic and environmental pressures, recycling of tungsten carbide has gained prominence as a cost-effective and sustainable alternative. This process allows for the recovery of valuable materials such as tungsten and cobalt, thereby reducing reliance on primary resource extraction and lowering overall production costs [10].

Despite these challenges, tungsten carbide remains indispensable in modern industry due to its unmatched hardness and resistance to wear and heat. The exceptional properties of tungsten carbide make it a superior material for a wide range of demanding applications. Its strength and durability make it the preferred choice for cutting tools, drills, and milling cutters used to machine both metallic and nonmetallic materials under extreme operating conditions [11]. These attributes enable tools to maintain performance over extended periods, thereby reducing operational costs and enhancing production efficiency across industries such as automotive, aerospace, mining, and construction [12–14]. Its applications also extend beyond cutting tools to high-technology sectors, underlining its strategic importance. Nevertheless, the associated health risks and sustainability concerns demand continuous attention and innovation to ensure its responsible and efficient use.

Tungsten carbide can be classified into several types based on its composition and intended application.

(i) Cutting and Machining Grades: These grades are primarily used in cutting tools, milling cutters, and drills due to their superior hardness and wear resistance. They typically contain a high proportion of tungsten carbide to maximize hardness and tool longevity.
(ii) Impact-resistant Grades: Designed for applications involving high mechanical shock, such as mining and drilling tools, these grades have a relatively higher cobalt content to enhance toughness and impact resistance.
(iii) Fine-grain Grades: These offer a well-balanced combination of hardness and toughness, making them ideal for applications requiring high wear resistance and precise surface finishes.

The classification of tungsten carbide grades is based on the ratio of hard phases to binder matrix, as well as the size of the tungsten carbide particles. Precise manufacturing techniques, such as careful control during sintering and component mixing, are essential to achieving desired material properties. Grades with higher tungsten carbide content exhibit superior hardness and wear resistance, while those with elevated cobalt content provide improved toughness and fracture resistance, though at the cost of some hardness. Furthermore, fine-grain structures enhance both strength and surface finish quality, making them particularly well-suited for precision applications [13].

The quality of tungsten carbide is influenced by factors such as grain size, the proportion of the cobalt matrix, and the presence of functional coatings. The microstructure plays a critical role in determining wear resistance and toughness. Fine-grain tungsten carbides offer enhanced wear resistance, making them ideal for precision tooling. The uniform distribution of fine tungsten carbide grains improves tool strength, particularly under high-abrasion conditions [7, 15]. On the other hand, modern cutting tools often incorporate functional and multilayer coatings, such as aluminum-titanium nitride and aluminum-chromium nitride, which provide exceptional wear resistance, thermal stability, and oxidation resistance. These properties are particularly crucial for high-performance cutting applications, including high-speed and lubrication-free machining. Physical Vapor Deposition (PVD) coatings, for example, have been shown to significantly enhance tool life and performance by providing a thermal barrier and reducing both abrasive and adhesive wear [16].

Tungsten carbide-based materials are classified into various grades based on microstructure and the relative proportions of tungsten carbide and cobalt, with the most common grades being K and P. Figure 3.3 shows these grades along with their subclassification groups. The K-grades, such as K10 and K20, are specifically designed for machining abrasive materials like cast iron and nonferrous metals. The K10 grade, in particular, strikes an optimal balance between hardness and toughness, making it ideal for continuous cutting applications due to its sharp edge retention and extended wear resistance [17]. In contrast, the P-grades are optimized

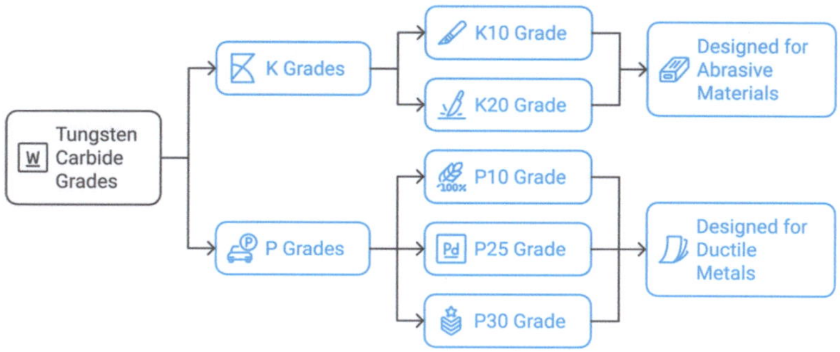

Fig. 3.3 Tungsten carbide grades

for machining steels and other ductile metals. These grades contain higher cobalt content, which enhances toughness compared to those designed for abrasive materials. Recent studies on P-grade tools, including P10, P25, and P30, have shown that variations in grain size and composition significantly influence wear resistance and crack resistance during turning and milling operations on carbon steels [18–22].

3.3 Texturing on Cutting Tools

The fundamentals of tool surface texturing lie in the strategic modification of cutting tool surfaces to optimize their interaction with the workpiece material during machining. Surface texturing involves creating microscale or nanoscale features, such as grooves, dimples, ridges, or structured arrays, on the tool surface to enhance its tribological, thermal, and mechanical performance. This modification is designed to address critical challenges in machining, including reducing friction, improving heat dissipation, increasing wear resistance, and facilitating efficient chip flow. These improvements are particularly important when machining difficult-to-cut materials. At the core of surface texturing is its ability to modify the tool–chip interface, a critical zone where intense mechanical and thermal stresses are concentrated. Textured patterns serve as microreservoirs for lubricants or cutting fluids, enhancing lubrication efficiency and reducing adhesion between the tool and the workpiece material, as shown in Fig. 3.4. This reduction in adhesion minimizes built-up edge (BUE) formation and lowers friction—both major contributors to tool wear and deterioration of surface quality. Furthermore, certain texture designs can improve chip control by guiding chip flow and promoting chip breakage, thereby preventing material clogging and ensuring a more stable cutting process [23].

From a thermal standpoint, textured surfaces contribute to the regulation of heat generation and dissipation during machining. By reducing the contact area and improving lubrication, these surfaces lower the thermal load on the cutting edge, thereby preventing excessive temperature rise that can lead to premature tool failure.

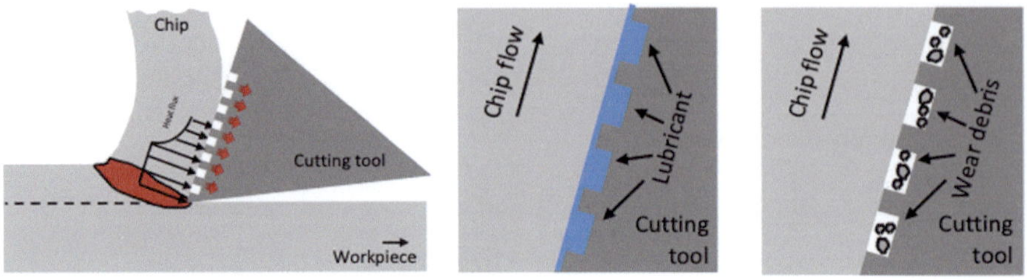

Fig. 3.4 Effects of textures on cutting tool surface

Moreover, surface texturing enhances wear resistance by redistributing mechanical stresses and mitigating the effects of abrasive and adhesive wear mechanisms. The precise implementation of these textures is made possible through advanced techniques such as laser surface texturing, electrochemical machining, and additive manufacturing, which allow for the creation of application-specific surface patterns. As a foundational innovation in cutting tool engineering, surface texturing offers a promising pathway to increased machining efficiency, extended tool life, and improved surface quality, particularly in high-performance and sustainability-focused manufacturing environments. Regarding the surface properties, texturing can significantly change the tribological behavior of cutting tools by improving heat dissipation and reducing cutting forces. Recent studies [24] showed that textured tools can reduce cutting forces by up to 21% and flank wear by 19%, thereby extending tool life. Furthermore, textured tools significantly reduce frictional effects and adhesive wear in high-speed and dry machining environments, where friction control is especially critical. For instance, the application of specific microtextures on carbide tools used for machining difficult-to-cut alloys has resulted in tool life improvements of up to 30% [25–27].

The effectiveness of microtexturing in extending tool life stems from its ability to optimize tool–material interaction, control adhesion, and enhance lubricant retention. Textured surfaces introduce functional features that reduce friction and wear by serving as microreservoirs for lubricants or cutting fluids. This localized lubrication lowers friction, reduces heat generation, and mitigates thermal and adhesive wear mechanisms [21]. Furthermore, microtextures promote more efficient heat dissipation by acting as thermal dispersion sites, thereby minimizing heat accumulation at the cutting edge and reducing diffusion and oxidation wear. This contributes to longer tool life. Moreover, texturing decreases the tendency of workpiece material to adhere to the tool surface, further limiting adhesive wear and improving the surface quality of machined components [22].

Given these advantages, tool surface texturing has emerged as one of the most advanced and innovative technologies for enhancing cutting tool performance and minimizing wear during machining operations. This technique involves the creation of microscopic patterns or engineered structures on the tool surface using methods such as laser engraving or micromachining. Its growing popularity is attributed to its ability to alter the tribological and thermal behavior of tools, specifically by reducing friction, minimizing wear, and improving cutting efficiency. Tool texturing primarily aims to modify the tribological interaction at the tool–workpiece interface by controlling surface contact. Patterns such as grooves, microholes, or porous features significantly influence friction reduction, heat dissipation, and the behavior of cutting fluids, thereby contributing to more stable and efficient machining processes [28].

3.4 Texturing Techniques and Patterns on Cutting Tools

Several techniques are available for modifying the surface of cutting tools, with laser texturing being one of the most widely used and effective methods. Laser texturing method employs laser beams to generate precise surface patterns on the tool surface. A laser functions as a transducer, converting electrical energy into a highly focused beam of monochromatic, coherent light. The key advantages of laser processing stem from the high alignment and energy concentration of laser beam, allowing for precise material removal. The laser texturing process operates primarily through ablation—a mechanism of rapid material vaporization—which minimizes melting and preserves surface integrity and dimensional accuracy [8, 22]. Laser machining enables the fabrication of highly precise features, including microholes as small as 315 μm, as well as intricate surface textures and engravings. It offers several advantages, including high processing speed, cost-effectiveness, precision, and negligible tool wear. However, despite these strengths, laser machining is not ideally suited for large-scale batch production due to its relatively lower throughput. Nevertheless, its versatility is notable, as it can be applied to a broad range of materials with varying thermal and mechanical properties [8]. Laser texturing is among the most extensively studied methods due to its versatility, precision, and controllability. In previous studies, researchers have investigated a wide range of laser parameters, such as power output, pulse duration, repetition rate, and scanning strategies, to tailor surface textures for specific materials and functional requirements. These parameters play a crucial role in determining the geometry, depth, and uniformity of the textures, which in turn influence the tribological and thermal performance of the cutting tools [20–22]. Figure 3.5 shows the schematic illustration of laser texturing process.

Chemical etching, also known as chemical milling, is another method for surface texturing that involves the selective removal of material from the surface using specific chemicals. The process is outlined in Fig. 3.6. One of the main advantages of chemical etching is that it does not affect the intrinsic mechanical properties of the material, making it a cost-effective solution for surface modification. However, its limitations include relatively slow processing times and reduced resistance to fatigue and surface roughness compared to other texturing methods [29].

Another surface texturing method is Electro Discharge Machining (EDM), which is widely used for creating microtextures on cutting tool surfaces. This process relies on the erosion of material through discrete electrical discharges between the electrode and the workpiece, which is submerged in a dielectric fluid [30, 31]. The electrical energy is converted into thermal energy, generating a plasma channel between the

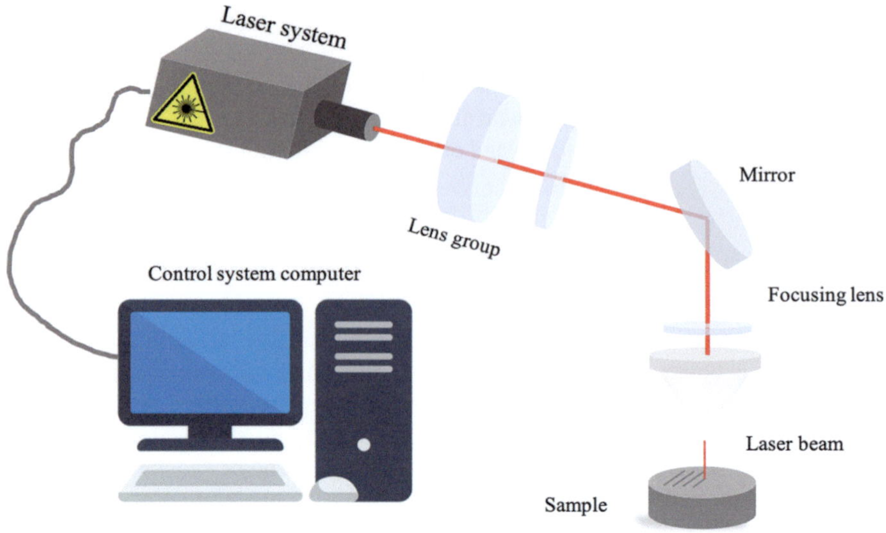

Fig. 3.5 Schematic illustration of laser texturing process

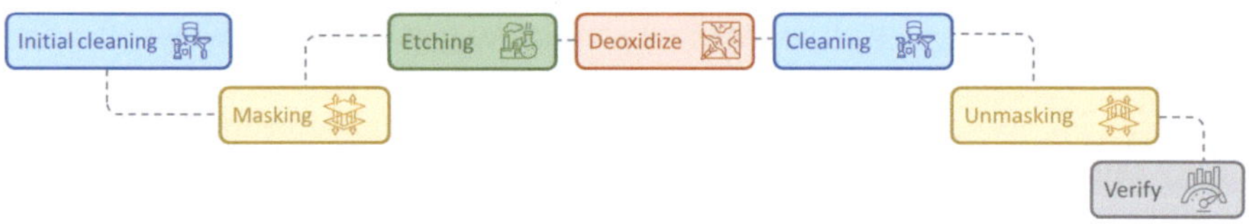

Fig. 3.6 Process flow in chemical texturing process

cathode and anode at temperatures ranging from 8,000 to 12,000 °C, and in some cases, as high as 20,000 °C. This extreme heat causes localized melting of the material at the surface of the workpiece. When the current is turned off, the plasma channel collapses, and the dielectric fluid flows through, removing the molten material in the form of microscopic debris. The material removal rate typically ranges from 2 to 400 mm^3/min, depending on the specific application. One of the key advantages of EDM is its high precision, as the shape of the electrode is replicated on the workpiece, resulting in accurate and fine textures on the tool surface.

Another widely used technique for creating textured surfaces is abrasive blasting. This process involves the impact of hard particles on the work surface to form the desired texture. One of the most notable methods in this category is Abrasive Jet Machining (AJM), which utilizes an ultrahigh-pressure water jet generated by a pump with pressures ranging from 3,000 to 6,000 bar. The process works by expelling water through a very small orifice (typically between 0.1 and 0.5 mm) at sufficient pressure to cut the material. The orifice is often made from precious stones, such as sapphire, ruby, or, less frequently, diamond. This creates a high-speed, narrow jet of water. The water is then mixed with abrasive particles in a chamber, which are carried at high velocity and directed through a nozzle to achieve the desired texture on the work surface [32–34].

The application of various texturing patterns on cutting tools marks a significant advancement in customizing tool performance for specific machining conditions and material types. Patterns such as grooves, dimples, ridges, and lattice structures are deliberately engineered and applied to critical tool surfaces, particularly the rake face and flank face, to enhance tool–workpiece–chip interactions. Each texture pattern offers distinct functional benefits based on its geometry, orientation, and dimensions. This allows for a highly tailored approach to mitigating common machining challenges, including excessive friction, heat buildup, chip adhesion, and accelerated tool wear. Groove patterns, for example, are often aligned with the direction of chip flow to promote smoother chip evacuation and reduce chip adhesion, an advantage particularly relevant when machining materials susceptible to BUE formation. Dimples, typically circular in shape, act as microreservoirs for lubricants or cutting fluids, thereby enhancing lubrication efficiency and minimizing friction and wear at the tool–chip interface. Ridge patterns contribute to improved wear resistance by redistributing mechanical stresses across the tool surface and mitigating abrasive and adhesive wear mechanisms that otherwise shorten tool life [35, 36].

A wide variety of texture pattern designs exist as shown in Fig. 3.7. The selection of a specific pattern depends on several factors, including the workpiece material, cutting parameters, and the required trade-off between tool life and surface quality. For instance, in high-speed machining of hard-to-cut materials such as titanium- or nickel-based superalloys, fine and densely spaced textures are often preferred to improve heat dissipation and protect the tool from thermal damage. In contrast, when machining softer materials, coarser and more widely spaced patterns may be sufficient to maintain effective chip control and lubrication. Advanced manufacturing techniques, such as laser surface texturing and chemical etching, allow for precise control over the dimensions and distribution of surface patterns, enabling performance optimization for specific machining applications. The deliberate selection and implementation of these texturing patterns represent a transformative advancement in cutting tool design, significantly enhancing tool efficiency, productivity, and sustainability. The primary goal of cutting tool texturing is to improve performance during machining by optimizing key aspects of the cutting process. This is achieved by introducing microscale or nanoscale surface features, strategically placed to influence friction, heat dissipation, wear resistance, and chip control [37–39].

As shown in Fig. 3.8, each texture pattern on a cutting tool can be defined by several geometric parameters: depth or height, diameter or width (depending on the pattern type), spacing between individual textures (pitch), and the concentration area, which refers to the proportion of the tool surface—particularly near the cutting edge—covered by the texture. Another critical factor influencing tool performance is the orientation of the texture. Textures can be aligned parallel or perpendicular to the main cutting edge, or oriented in the direction of chip flow. The choice of orientation significantly affects chip evacuation, friction reduction, and overall cutting efficiency.

Various pattern geometries play an important role in the performance of textured cutting tools, with factors such as shape, size, and surface density significantly

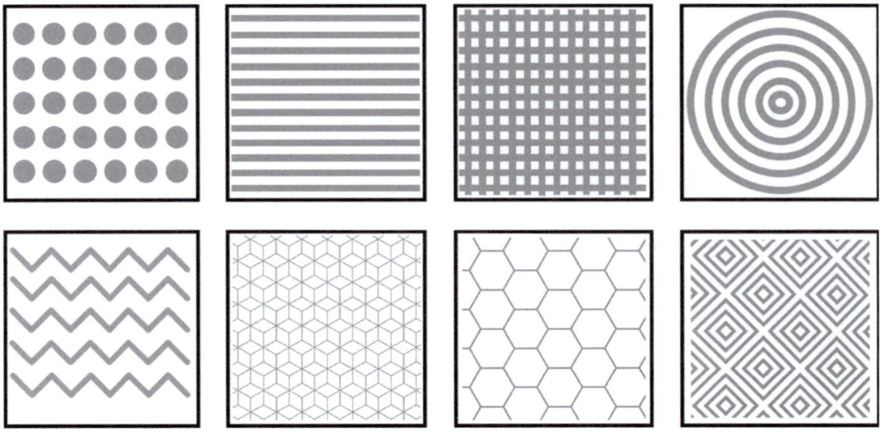

Fig. 3.7 Various texture pattern designs for cutting tools

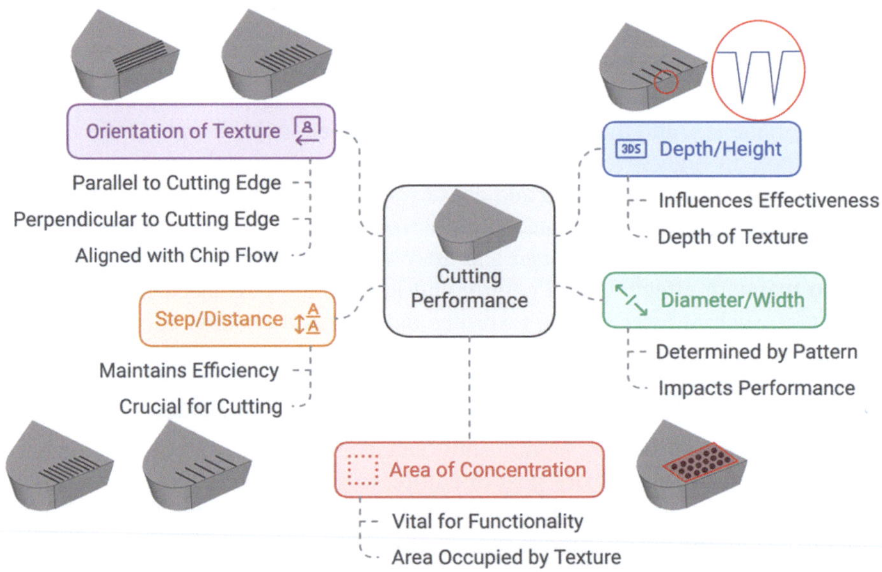

Fig. 3.8 Geometric and dimensional texture characterization factors

influencing machining outcomes. To enhance machining quality over conventional methods, optimizing these parameters is essential. Among the most effective and widely studied patterns are dimples, typically hemispherical in shape and easily identifiable. While circular dimples are the most common, other geometries such as cylindrical and pyramidal forms have also been employed. These textures have proven particularly effective when machining low-machinability materials like heat-resistant alloys, demonstrating improved tool performance and wear resistance [40].

The effectiveness of dimple textures depends significantly on their dimensions, depth, and area of application. Zhang et al. [41] investigated the influence of microtextures on the wear mechanisms of cemented carbide (WC-TiC/Co) tools during the machining of AISI 1045 steel (C: 0.42–0.5%), a material known for its hardness and machining difficulty. In their study, dimples were applied on the rake face of the tool in a linear symmetrical pattern. The optimal performance was achieved with dimples measuring 170 μm in diameter, 7 μm in depth, and covering 20% of the surface area. These parameters significantly reduced shear strength, frictional forces, and the temperature at the tool–chip interface, thereby enhancing the frictional and wear resistance characteristics of the cutting tool. Figure 3.9 shows the resultant forces and average friction coefficients for the machining operations with the nontextured and textured cutting tools.

Another notable texture type is the groove pattern, which is characterized by its elongated, channel-like structure. Similar to dimples, groove patterns can exhibit a variety of cross-sectional shapes, but they differ primarily in their directional alignment. The effectiveness of groove textures is largely influenced by their orientation relative to the cutting edge or the direction of chip flow. Most studies have investigated groove orientations on the rake face in three common configurations: perpendicular, parallel, or diagonal to the cutting edge. However, some researchers have also explored aligning the grooves in accordance with the chip flow direction, which may offer additional benefits in chip evacuation and friction reduction. Figure 3.10 shows linear groove patterns on a cutting tool. Enomoto et al. [42] demonstrated that linear groove textures aligned parallel to the cutting edge significantly reduce chip adhesion, thereby lowering friction and adhesive wear. This orientation provided superior performance compared to grooves placed perpendicularly. In their study, microtextures with depths ranging from 100 to 150 μm and a groove spacing of 700 μm applied to the rake face of Diamond-like Carbon (DLC)-coated cemented carbide tools enhanced the antiadhesion properties during the machining of aluminum alloy

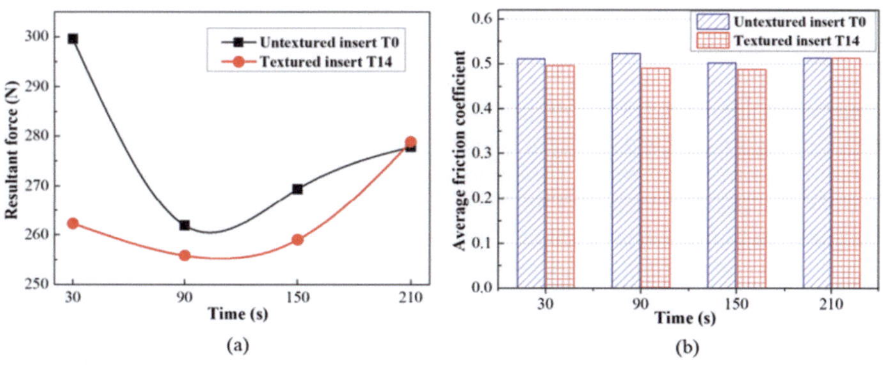

Fig. 3.9 **a** Resultant forces and **b** average friction coefficients for machining operations with nontextured and textured cutting tools [41]

Fig. 3.10 Linear groove patterns developed by laser texturing on a carbide cutting tool

AA5052. In addition, under lubricated conditions, adhesion during cutting was effectively reduced by enhancing the retention of cutting fluid on the tool surface. Tools with nano-/microgrooves aligned parallel to the main cutting edge exhibited superior antiadhesive properties compared to other samples.

While straight grooves are the most commonly used, other types of textures such as sine wave grooves, arc grooves, zigzag grooves, rectangular textures, and other segmented patterns can also be employed [43]. These groove patterns have proven to be highly effective, especially when combined with solid lubricants. This combination allows for a significant increase in cutting speed without causing additional damage, while also reducing cutting forces. In contrast to groove patterns, there are positive patterns in the form of protrusions. These protrusions, often circular, have shown significant efficiency improvements in machining processes. For example, studies have demonstrated their effectiveness when machining stainless steel with textures ranging from 150 to 310 μm in size, texture densities of 5–10%, and texture heights between 100 and 300 μm, achieving tool life improvements of up to 110%. Similarly, for materials like Ti6Al4V, protrusions as small as 10 μm have been shown to reduce cutting temperature by 25% and the coefficient of friction by 11.1%, compared to conventional tooling. This reduction in friction results in significant decreases in cutting force, thrust force, and overall wear.

The purpose of texturing in cutting tools goes beyond conventional tribological improvements, addressing critical aspects such as coating adhesion, wear control, friction reduction, and chip management. By precisely modifying tool surfaces to create microscale or nanoscale patterns, texturing significantly enhances the tool performance under demanding machining conditions. One of the key applications of these textures is to improve the adhesion of coatings commonly applied to cutting tools, such as titanium nitride (TiN), titanium aluminum nitride (TiAlN), or DLC. The textured surfaces increase the contact area and provide mechanical interlocking sites for coatings, which enhances their bond strength and durability. This is particularly vital in high-performance machining environments, where coatings must withstand extreme thermal and mechanical stresses [44]. Figure 3.11 shows the SEM images of coated samples with dimple and groove textures. Due to the intricate surface topographies produced by the texturing process, coating adhesion is enhanced compared to nontextured samples. In addition to improving coating adhesion, texturing plays a pivotal role in wear control by redistributing mechanical stresses and mitigating prevalent wear mechanisms such as adhesion, abrasion, and crater wear. The patterned features function as stress-relief zones that interrupt or deflect the progression of wear-related damage, thereby significantly extending the operational life of the tool. Moreover, these textures reduce the coefficient of friction at both the tool–chip and tool–workpiece interfaces by enhancing lubrication performance. Patterns such as grooves and dimples act as microreservoirs for cutting fluids or as traps for wear debris, helping to maintain a low-friction environment that minimizes heat generation and material adhesion.

Another essential function of tool texturing is effective chip control. Textured surfaces can influence chip flow direction, promote chip breaking, and help prevent issues such as chip clogging or entanglement—factors that are critical for maintaining machining stability and achieving high surface quality. By enabling smoother chip evacuation and minimizing the formation of BUE, texturing contributes to prolonged tool life and improved finish on the machined surface. As an integrated solution, surface texturing empowers cutting tools to meet the increasing performance and reliability demands of advanced manufacturing, ultimately enhancing efficiency, durability, and overall machining performance [45]. Surface texturing on cutting tools also regulates heat generation and dissipation during machining. By reducing friction and improving contact conditions at the tool–chip interface, textured tools decrease the thermal load on the cutting edge. Certain texture designs can enhance heat conduction or act as microchannels for thermal dispersion, thereby preventing excessive temperature buildup that could lead to thermal softening or damage to both the tool and

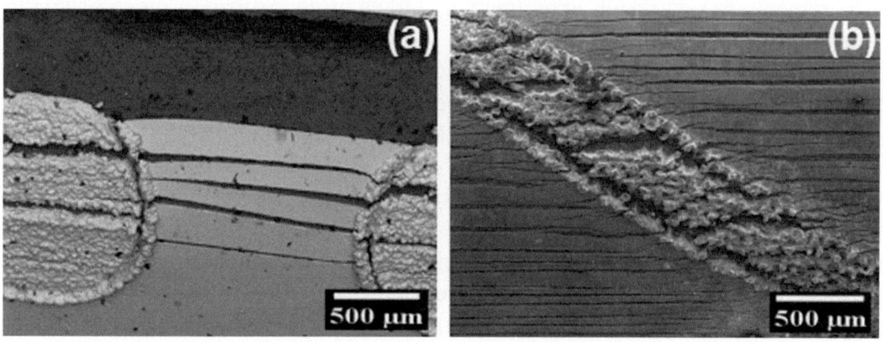

Fig. 3.11 SEM images of coated samples with **a** dimple and **b** groove textures [44]

workpiece. This thermal management is particularly critical in high-speed and dry machining operations, where the lack of coolant intensifies heat-related challenges. The combined effects of reduced friction, enhanced wear resistance, and controlled heat generation contribute to a significant extension of tool life. Increased tool longevity reduces the need for frequent tool changes, thereby minimizing downtime and lowering operational costs. This, in turn, boosts overall productivity. Furthermore, the consistent performance of textured tools ensures improved surface finish and dimensional accuracy in machined components, ultimately enhancing product quality [37, 38].

The experimental evaluation of cutting tool texturing is essential for understanding its impact on machining performance and for optimizing its application to specific materials and processes. This evaluation systematically examines key variables such as texture geometry and texture density. In addition, it involves the analysis and characterization of textured surfaces using precise measurement techniques. These parameters are critically assessed to determine their influence on tool wear, friction reduction, chip formation, heat dissipation, and overall machining efficiency. The resulting insights help clarify the mechanisms by which surface texturing enhances the performance and durability of cutting tools [37]. Surface roughness is a critical factor in determining the nature of interaction between the cutting tool and the workpiece. An effective texture design is needed for minimizing contact area to reduce friction, while avoiding excessive roughness that could introduce stress concentrations and accelerate tool wear or failure. Experimental setups are designed to maintain strict control over surface roughness, typically using advanced manufacturing techniques such as laser surface texturing or chemical etching. These methods enable high precision and repeatability, ensuring consistency in the geometry and distribution of textured patterns. Roughness measurements are particularly useful for evaluating wear on textured or modified tool surfaces.

The texture density, defined as the ratio of the textured area to the total surface area, plays a crucial role in determining the performance of textured cutting tools. Higher texture densities can enhance lubrication retention and wear resistance by increasing the number of functional microfeatures. However, excessive density may compromise the structural integrity of the cutting edge or negatively influence chip flow dynamics. As a result, experimental studies often evaluate a range of texture densities to determine the optimal configuration that balances performance gains with mechanical stability. Dan et al. [46] fabricated chevron-patterned microtextures on cemented carbide tools, varying the texture density as shown in Fig. 3.12. According to the friction coefficient results presented in Fig. 3.13, friction gradually decreased as the texture density increased, reaching a minimum at a density of 9.5%. Beyond this threshold, however, the friction coefficient rose sharply. Although higher texture densities introduce more microdimples capable of generating hydrodynamic pressure, the overall load capacity of the lubricating film and the pressure contribution from each unit become limited. Moreover, when the spacing between microcells becomes too small, fluid tends to flow quickly from the high-pressure zone to the adjacent low-pressure zone, reducing the stability of the lubrication regime. Therefore, a texture density of 9.5% appears to be optimal, balancing microcell distribution

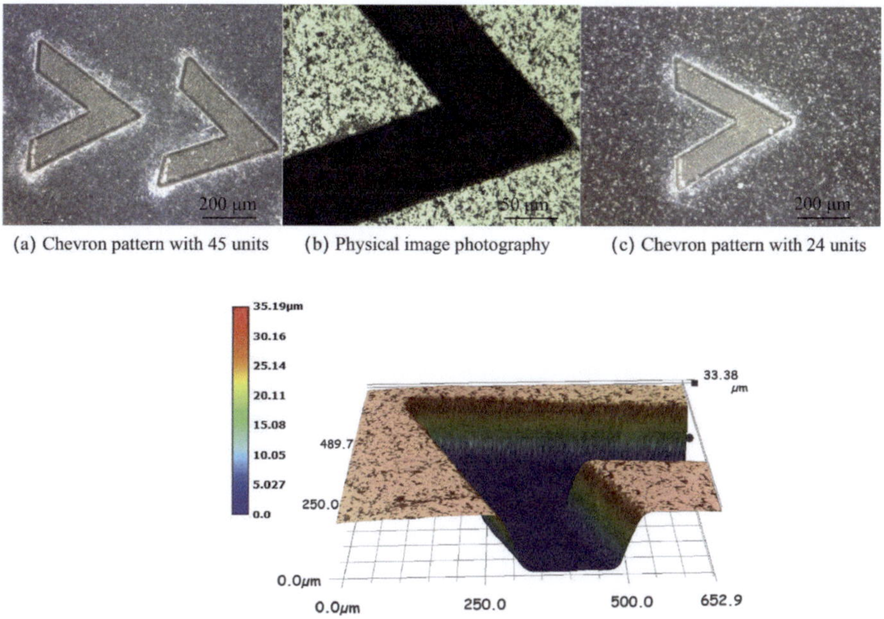

(a) Chevron pattern with 45 units (b) Physical image photography (c) Chevron pattern with 24 units

(d) 3D display of micro-texture composition

Fig. 3.12 Chevron patterns textured on cemented carbide tools [46]

and hydrodynamic pressure generation to effectively minimize friction. Tribological testing methods, such as pin-on-disc and ball-on-flat tests, are also employed to evaluate the friction and wear behavior of textured surfaces under controlled laboratory conditions. These tests simulate the contact mechanics between the cutting tool and workpiece material, offering critical insights into the effectiveness of different texture geometries. The data obtained from these tests are instrumental in optimizing surface textures for specific machining applications. Pin-on-disc testing conducted by Dan et al. [46] revealed the combined influence of rotational speed and applied load on the friction coefficient of a chevron-patterned surface with a 9.5% texture density, as shown in Fig. 3.14.

In machining trials, the performance of textured tools is evaluated under real cutting conditions, with parameters such as cutting speed, feed rate, depth of cut, and lubrication strategy systematically controlled. Key performance indicators, including tool wear rate, cutting forces, surface finish of the machined part, and chip morphology, are measured to assess the effectiveness of surface textures. Moreover, thermal imaging and infrared thermography are frequently employed to monitor heat generation and distribution during the cutting process, offering valuable insights into the thermal management capabilities of the textured tools. Ze et al. [47] investigated heat generation during dry machining of Ti6Al4V alloys by evaluating three types of cemented carbide cutting tools: a conventional nontextured tool (CT), a tool with elliptic textures on the rake face (STT-R), and a tool with linear

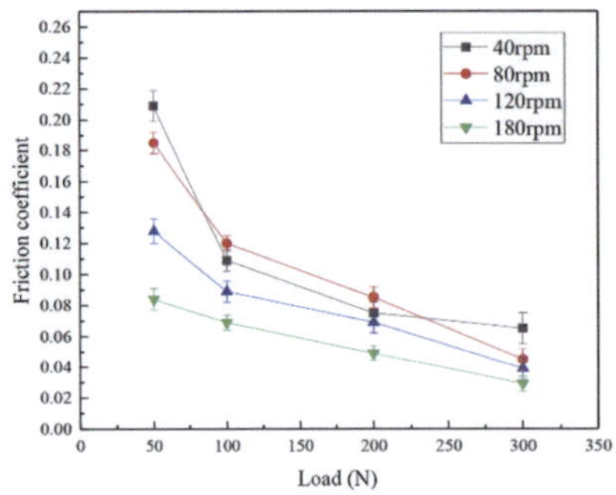

(a) the sample with 5.05% texture density

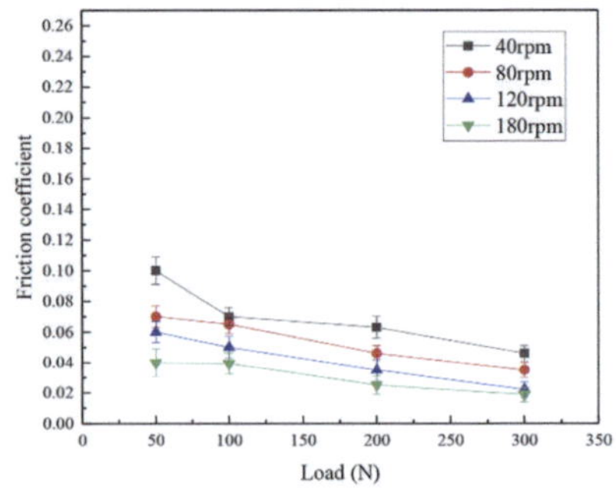

(b) the sample with 9.5% texture density

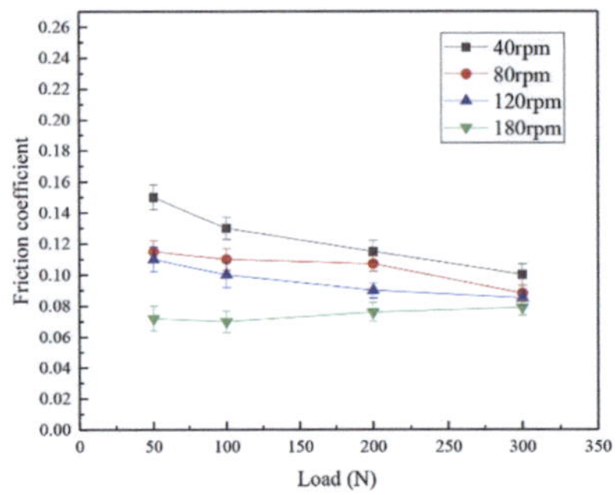

(c) the sample with 13.02% texture density

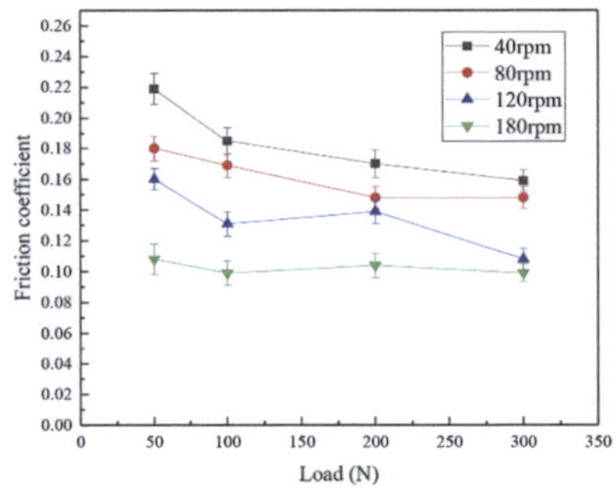

(d) the sample with 15.2% texture density

Fig. 3.13 Friction coefficients for various texture densities [46]

grooves on the flank face (STT-F). Figure 3.15a shows the heat distribution during machining with the STT-R tool, as captured by an infrared thermal imaging system. The maximum cutting temperatures corresponding to different cutting speeds are presented in Fig. 3.15b, providing a comparative analysis of thermal performance among the tool configurations. In addition to experimental efforts, the integration of computational simulations, particularly finite element analysis (FEA), with experimental studies provides a powerful, complementary approach for evaluating the performance of textured surfaces. These simulations can predict critical parameters such as stress distribution, heat flow, and material removal mechanisms, offering valuable insights that support the interpretation of experimental findings. Moreover, computational modeling facilitates the systematic optimization of texture designs by enabling the evaluation of numerous configurations and machining scenarios without the need for extensive physical testing. Saraf et al. [48] developed micro-textured cemented carbide cutting tools by systematically varying the diameter and spacing of surface protrusions, as shown in Fig. 3.16. To evaluate the impact of these texture parameters, they conducted machining operations on Ti6Al4V alloy under controlled conditions: a cutting speed of 110 m/min, a feed rate of 0.3 mm/rev, and a

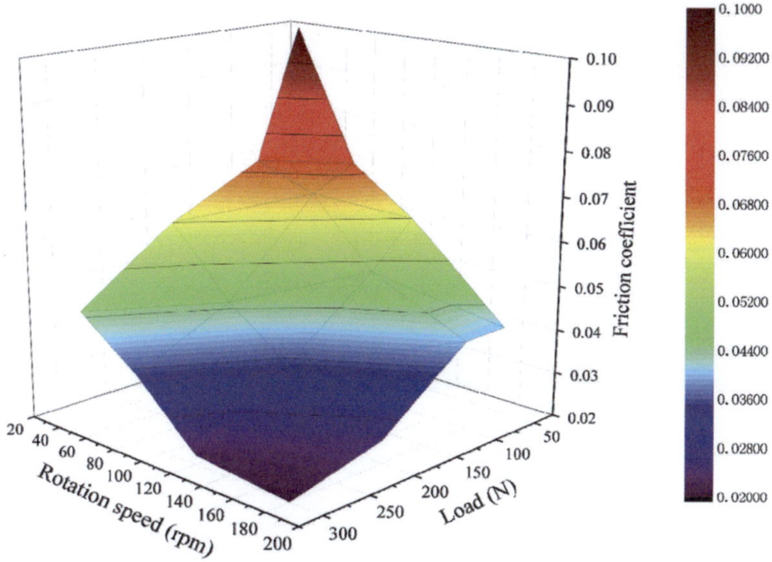

Fig. 3.14 Friction coefficients with respect to rotational speed and applied load for a chevron-patterned surface with texture density of 9.5% [46]

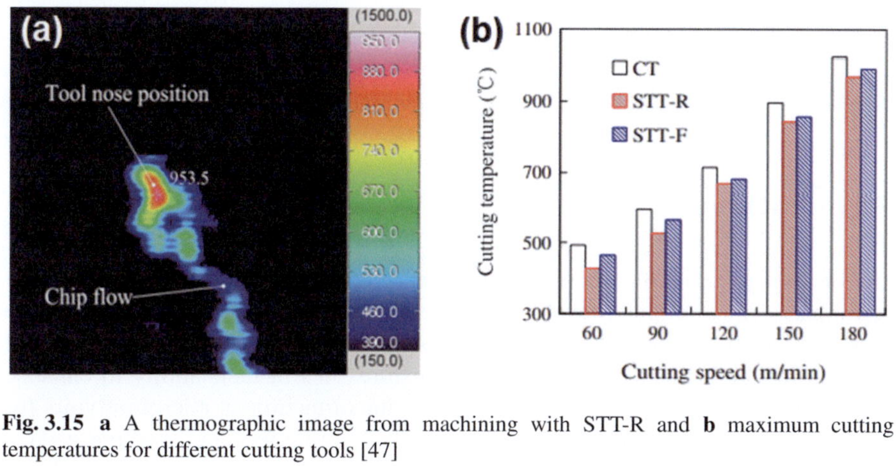

Fig. 3.15 **a** A thermographic image from machining with STT-R and **b** maximum cutting temperatures for different cutting tools [47]

depth of cut of 0.5 mm. FEA simulations were employed to predict heat distribution and investigate the influence of texture geometry on thermal behavior. These simulations were validated against experimental results, demonstrating good agreement. Figure 3.17 presents the temperature contour plots derived from the FEA simulations, highlighting the effect of protrusion diameter and spacing on heat dissipation during cutting.

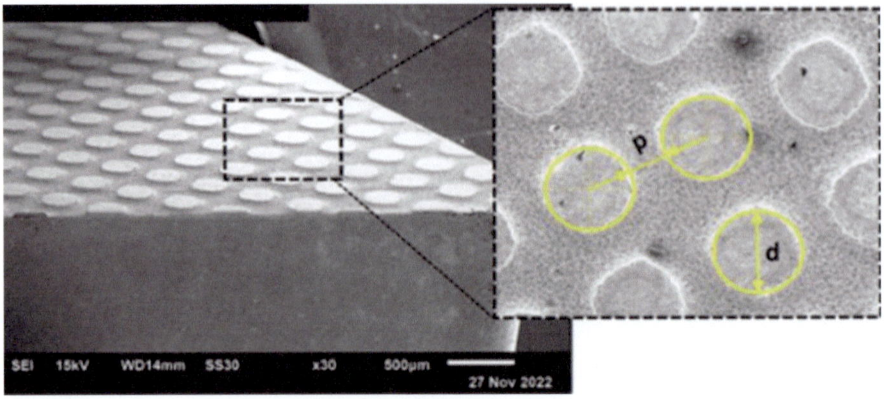

Fig. 3.16 Microtextured patterns with different diameters and spacings of protrusions on rake face of cutting tool [48]

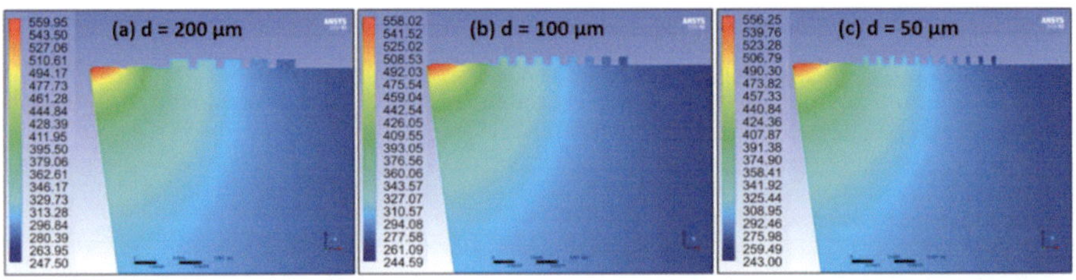

Fig. 3.17 Temperature contours for textured tools with varying protrusion diameters and spacings [48]

3.5 Conclusion

Microtexturing on carbide cutting tools is an effective strategy for enhancing the performance of material removal operations. A variety of microtexture designs have been applied to improve key process outcomes, including surface finish and tool life. One of the primary benefits of microtextures lies in their ability to enhance lubrication within the cutting zone. These textures function as microreservoirs for lubricants, thereby reducing friction between the cutting tool and workpiece. This results in smoother chip flow along the tool surface and contributes to reduced tool wear. Furthermore, microtexturing contributes to lowering cutting temperatures by increasing the tool surface area and facilitating improved penetration of coolants into the tool–workpiece interface. Furthermore, microtextures can serve as chip breakers, disrupting continuous chip formation that would otherwise degrade the surface quality of the machined workpiece.

Acknowledgements This work was supported by MCIN/AEI/10.13039/501100011033 and ERDF/EU [grants PID2022-138872OB-I00 and EQC2019-005674-P], and the University of Cadiz [grant 2023-046/PU/EPIF-FPI-CT/CP].

References

1. Okamoto S, Nakazono Y, Otsuka K, Shimoitani Y, Takada J. Mechanical properties of WC/Co cemented carbide with larger WC grain size. Mater Charact 2005, 55 (4–5), 281–287.
2. Zeng K, Wu X, Jiang F, Shen J, Zhu L, Li L. A comprehensive review on the cutting and abrasive machining of cemented carbide materials. Journal of Manufacturing Processes 2023, 108, 335–358.
3. Bouzakis K, Michailidis N, Skordaris G, Bouzakis E, Biermann D, M´Saoubi R. Cutting with coated tools: Coating technologies, characterization methods and performance optimization. CIRP Annals 2012, 61, 703–723.
4. Phokobye S N, Desai D A, Tlhabadira I, Sadiku E M, Mutombo K. Comparative investigation and optimization of cutting tools performance during milling machining of titanium alloy (Ti6Al4V) using response surface methodology. Int J Adv Manuf Technol 2024, 131, 3565–3577.
5. Ranjan P, Hiremath S S. Role of textured tool in improving machining performance: A review. Journal of Manufacturing Processes 2019, 43, 47–73.
6. Machado A R, Da Silva L R R, De Souza F C R, Davis R, Pereira L C, Sales W F, Ezugwu E O. State of the art of tool texturing in machining. Journal of Materials Processing Technology 2021, 293, 117096.
7. García J, Ciprés V C, Blomqvist A, Kaplan B, Cemented carbide microstructures: a review. Int J Refract Metals Hard Mater 2019, 80, 40–68.
8. Šubić J, Slokar B L, Selanec M, Erman Ž. Effect of hard metal production on the environment. Holistic approach to environment 2022 12, 3, 102–109.
9. Fischbein A, Luo J, Solomon S, Horowitz S, Hailoo W, Miller A. Clinical findings among hard metal workers. British Journal of Industrial Medicine 1992, 49, 17–24.
10. Kulynych V, Shapoval A, Drahobetskyi V. Hard Alloys Recycling as a Promising Direction of Technological Equipment for Machine-Building Production. Materials Science Forum 2022, 1052, 423–428.
11. Cugell DW. The hard metal diseases. Clin Chest Med. 1992; 13, 269–79.
12. Amado J M, Tobar M, Alvarez J C, Lamas J, Yáñez A. Laser cladding of tungsten carbides (Spherotene®) hardfacing alloys for the mining and mineral industry. Applied Surface Science 2009, 255, 10, 5553–5556.
13. Mel'nichenko AS, Kudrya A, Akhmedova T S, Sokolovskaya E A. Predicting the Risk of Destruction of Hard-Facing Alloys Based on the Morphology of Their Structure. Metallurgist 2018, 61, 1130–1134.
14. Fang S, Salán N, Pauly C, Llanes L, Mücklich F. Critical Assessment of Two-Dimensional Methods for the Microstructural Characterization of Cemented Carbides. Metals 2022, 12(11), 1882.
15. Lengauer W, Dreyer K. Functionally graded hardmetals. Journal of Alloys and Compounds 2002, 338, 194–21.
16. Kalss W, Reiter A, Derflinger V, Gey C, Endrino J. Modern coatings in high performance cutting applications. International Journal of Refractory Metals & Hard Materials 2006, 24, 399–404.
17. Rao C, Chandra P, Kiran R, Asha P. Influence of machining parameters on cutting tool life while machining aluminum alloy fly ash composite. IOP Conference Series: Materials Science and Engineering 2016, 149, 012157.

18. Okusa K, Kitagawa R, Yamamoto E. On the Cutting Performance and Metallurgical Properties of Carbide Tools. Journal of The Japan Society for Precision Engineering 1976, 42, 113–119.
19. Kwon H. Improvement of mechanical properties of hard metals for cutting tools. Ceramist 2023, 26(3), 327–336.
20. Ibrahim G, Haron C, Ghani J. Tool Wear Mechanism in Continuous Cutting of Difficult-to-Cut Material under Dry Machining. Advanced Materials Research 2010, 126–128, 195–201.
21. Obikawa T, Kamio A, Takaoka H, Osada A, Micro-texture at the coated tool face for high performance cutting, International Journal of Machine Tools and Manufacture 2011, 51, 12, 966–972.
22. Sugihara T, Enomoto T. Development of a cutting tool with a nano/micro-textured surface—Improvement of anti-adhesive effect by considering the texture patterns, Precision Engineering 2009, 33, 4, 425–429.
23. Sristi N A, Zaman P B. A review of textured cutting tools' impact on machining performance from a tribological perspective. The International Journal of Advanced Manufacturing Technology 2024, 133, 4023–4057.
24. Elias J V, Venkatesh P, Lawrence K, Mathew J. Tool texturing for micro-turning applications – an approach using mechanical micro indentation. Materials and Manufacturing Processes 2020, 36, 84–93.
25. Jiang W. Bio-inspired self-sharpening cutting tool surface for finish hard turning of steel. CIRP Annals 2014, 63, 517–520.
26. Özel T, Biermann D, Enomoto T, Mativenga P. Structured and textured cutting tool surfaces for machining applications. CIRP Annals 2021, 70, 2, 495–518.
27. Roushan A, Chetan. Influence of laser parameters on the machining performance of textured cutting tools. Optics & Laser Technology 2023, 165, 109569.
28. Gachot C, Rosenkranz A, Hsu S, Costa H. A critical assessment of surface texturing for friction and wear improvement. Wear 2017, 372, 21–41.
29. Astakhov V. The assessment of cutting tool wear. International Journal of Machine Tools & Manufacture 2004, 44, 637–647.
30. Khandizod R, Varghese V, Mujumdar S. Electric Discharge Assisted Surface Texturing of Stainless Steel 304. Procedia CIRP 2022, 108, 670–674.
31. Guu Y H, Ti-Kuang Hou M. Effect of machining parameters on surface textures in EDM of Fe-Mn-Al alloy. Materials Science and Engineering A 2007, 466, 61–67.
32. Tshimanga N L, Combrink G A, Kalenga M W. Surface morphology characterization of grade 304L stainless steel after abrasive blasting. Materials Today: Proceedings 2021, 38, 544–548.
33. Reddy P S, Kumar P S, Prasad D V S S S V, Rani B S, Gupta C N, Das A D, Subbiah R. Effect of parameters and surface analysis on Eglin steel by shot blasting method. Materials Today: Proceedings 2023, 72, 2833–2836.
34. Stöckel S, Lia Z, Gröger S. Experimental study on the influence of blasting time on the surface topography of machined steel samples. Procedia CIRP 2024, 123, 77–82.
35. Gimeno S, Mescheder H, Quintana I, Gasion A, Arias-Egido E, Carbonell A, Mallo C, Miguel I, Paredes J, Zalakain I. Effect of different laser texturing patterns on rolling contact surface and its tribological & fatigue life behavior on 100Cr6 bearing steel. Wear 2023, 522, 204717.
36. Soni D L, Jagadish. Nature-inspired texture pattern for cutting tool tribological surface modification: A state of art. Materials Today: Proceedings 2022, 60, 1353–1357.
37. Ranjan P, Hiremath S S. Role of textured tool in improving machining performance: A review. Journal of Manufacturing Processes 2019, 43A 47–73.
38. Arslan A, Masjuki H H, Kalam M A, Varman M, Mufti R A, Morasof M H, Khuong L S, Quazi M M. Surface Texture Manufacturing Techniques and Tribological Effect of Surface Texturing on Cutting Tool Performance: A Review. Critical Reviews in Solid State and Materials Sciences 2016, 41(6), 447–481.
39. Li Q, Pan C, Jiao Y, Hu K. Investigation on Cutting Performance of Micro-Textured Cutting Tools. Micromachines 2019, 10(6) 352.
40. Rajurkar A, Chinchanikar S. Performance study of dimple and channel textured tools during turning of Inconel-718. Materials Today Proceedings 2021, 46 (17), 8347–8351.

41. Zhang J, Yang H, Chen S, Tang H. Study on the influence of micro-textures on wear mechanism of cemented carbide tools. The International Journal of Advanced Manufacturing Technology 2020, 108, 1701–1712.
42. Enomoto T, Sugihara T, Yukinaga S, Hirose K, Satake U. Highly wear-resistant cutting tools with textured surfaces in steel cutting. CIRP Annals 2012, 61, 1, 571–574.
43. Santana T, De Rossi W, Barbosa PA, Bertolete M. Performance of cutting-tool patterns textured via ultrashort laser pulses in the turning of martensitic stainless steel under dry and lubricated conditions. Proceedings of the Institution of Mechanical Engineers Part B Journal of Engineering Manufacture 2023, 238, 4.
44. Heidarinejad A, Ashrafizadeh F. Influence of surface texture and coating thickness on adhesion of nickel plated coatings to aluminium substrate. Journal of Manufacturing Processes 2024, 120, 435–448.
45. Aoki T, Sencer B, Shamoto E, Suzuki N, Koide T. Development of a high-performance chip-guiding turning process—tool design and chip flow control. The International Journal of Advanced Manufacturing Technology 2016, 85, 791–805.
46. Dan L, Xuefeng Y, Chongyang L, Jian C, Shouren W, Yanjun W. Tribological characteristics of a cemented carbide friction surface with chevron pattern micro-texture based on different texture density. Tribology International 2020, 142, 106016.
47. Ze W, Jianxin D, Yang C, Youqiang X, Jun Z. Performance of the self-lubricating textured tools in dry cutting of Ti-6Al-4V. The International Journal of Advanced Manufacturing Technology 2012, 62, 943–951.
48. Saraf G, Sharma G, Kumar R, Nirala CK. Experimental and numerical investigation on heat dissipation capability of micro-pillar textured cutting tools. Scientific Reports 2025, 15, 12282.

Chapter 4
Micro-electrical Discharge Machining for a Difficult-to-Cut Aerospace Alloy

Muthuramalingam Thangaraj, Geethapriyan Thangamani, and Panagiotis Karmiris-Obratański

Abstract This study investigates the effects of key micro-electrical discharge machining (micro-EDM) process parameters on the recast layer characteristics during the micromachining of Inconel 718, a nickel-based alloy widely used in aerospace applications. The influence of open circuit voltage, peak current, and duty factor was systematically investigated with respect to their impact on average recast layer thickness (AVLT), a critical indicator of surface integrity. In addition, the role of pulse generator type was evaluated by comparing a conventional transistor pulse generator (TPG) with a modified ISO pulse generator (IPG). While the TPG produces irregular energy discharges that contribute to uneven crater formation and a thicker recast layer, the IPG enables uniform energy delivery, resulting in more stable material removal. The findings contribute to optimizing micro-EDM performance for high-precision machining of difficult-to-cut aerospace materials.

Keywords Micro-electrical discharge machining · Noncontact machining · Difficult-to-cut materials · Aerospace alloys · Recast layer

M. Thangaraj (✉)
Department of Mechatronics Engineering, SRM Institute of Science and Technology, Kattankulathur, Chengalpattu, Tamil Nadu, India
e-mail: muthurat@srmist.edu.in

G. Thangamani
Laboratório Department of Applied Science and Technology (DISAT), Politecnico di Torino, Turin, Italy

P. Karmiris-Obratański
Advanced Manufacturing Laboratory, Department of Manufacturing Systems, Faculty of Mechanical Engineering and Robotics, AGH University of Krakow, Krakow, Poland

S. Gürgen (ed.), *Micromanufacturing Operations*,
https://doi.org/10.1007/978-3-032-04193-7_4

4.1 Introduction

Recent advancements in aerospace, automotive, and nuclear engineering have driven demand for high-performance materials characterized by exceptional strength, hardness, and thermal resistance. These materials—often referred to as difficult-to-machine alloys—pose significant challenges to conventional machining techniques due to their low machinability and high tool wear. In response, nontraditional machining methods have gained prominence, with micro-electrical discharge machining (micro-EDM) emerging as a leading technology for precision fabrication at the micro-scale.

Micro-EDM is a derivative of conventional EDM tailored for micromachining applications [1]. Like its macroscale counterpart, micro-EDM removes material through a series of controlled electrical discharges between a microelectrode and a conductive workpiece, eliminating the need for direct mechanical contact. This noncontact nature of material removal eliminates cutting forces, reduces tool wear, and enables machining of ultrahard materials [2–4]. However, micro-EDM distinguishes itself by operating at much lower discharge energies, often in the order of microjoules, and using ultrafine electrodes—thereby enabling the fabrication of features with sub-millimeter dimensions and micron-level accuracy. The micro-EDM process offers distinct advantages in machining high-strength aerospace alloys, such as Inconel 718, titanium alloys, and other nickel-based superalloys. These materials, while exhibiting excellent performance in high-temperature and high-load environments, are notoriously difficult to machine using conventional methods due to their high melting points, work-hardening tendencies, and poor thermal conductivity [5, 6]. Micro-EDM overcomes these barriers by enabling the creation of microholes, microgrooves, and complex cavities with minimal distortion and tight dimensional control. This makes it particularly suitable for manufacturing high-precision components such as fuel injector nozzles, turbine blade cooling channels, microvalves, and sensor housings used in critical aerospace applications [7–9].

Despite its advantages, micro-EDM still faces significant challenges related to surface integrity, which remains a critical determinant of component performance, particularly in fatigue and stress-sensitive aerospace environments. Each electrical discharge in micro-EDM results in localized melting and vaporization of the workpiece surface. Upon rapid solidification, a recast layer, also known as a white layer, is formed [10, 11]. Although the scale of this layer is reduced compared to conventional EDM, it can still exhibit micropores, thermal cracks, and residual stresses, all of which degrade the surface and subsurface quality of the machined part. These imperfections can act as stress concentrators, reducing fatigue life and potentially leading to premature failure in service. The formation of residual stress and microcracks is closely linked to the random and transient nature of the discharge energy [12–14]. In micro-EDM, where the process operates within narrow parameter windows and is often applied to miniaturized geometries, achieving uniform and stable spark energy is even more critical. Irregular energy input can lead to excessive heat accumulation, uncontrolled material ejection, and uneven crater formation—all of which compromise surface finish and dimensional precision [15–18].

To address these concerns, advanced pulse control technologies are being developed. Among them, modified pulse generators, such as the ISO energy pulse generator (IPG), offer enhanced control over discharge energy by generating consistent, low-energy pulses. These generators are designed to minimize thermal gradients and arcing effects, leading to more uniform material erosion and significant reductions in recast layer thickness, surface roughness, and microcrack density [19]. By enabling arcing-free, stable discharges, IPGs improve machining efficiency and prolong tool life—critical factors for cost-effective micromanufacturing in aerospace sectors. Moreover, attention must also be paid to the heat-affected zone (HAZ) below the machined surface [20]. In ferrous and nickel-based alloys, the elevated temperatures during machining can induce solid-state phase transformations, which contribute to the buildup of residual stresses. While these stresses may not always result in visible cracks, they can significantly affect the mechanical stability and fatigue resistance of the component under cyclic loading conditions—especially in high-speed, high-temperature aerospace environments.

In this context, micro-EDM presents a highly promising solution for micromachining difficult-to-cut aerospace materials, provided that adequate control is exercised over the discharge process. Achieving high-quality surfaces with minimal thermal damage demands not only optimal selection of process parameters—such as open circuit voltage, peak current, and duty factor—but also the incorporation of pulse generator technologies capable of delivering precise and reproducible discharge energy [21–23]. Previous studies, such as those by Muthuramalingam et al. [24], have shown that the use of IPG can significantly improve the quality of the machined surface by reducing recast layer formation and enhancing surface finish. However, comprehensive investigations are still required to quantify the exact benefits of IPGs across various process settings and material types.

This study aims to experimentally investigate the influence of micro-EDM process parameters on the formation and characteristics of the recast layer during the micromachining of Inconel 718, a nickel-based aerospace alloy. To enhance the micro-EDM process, three essential electrical parameters—open circuit voltage, peak current, and duty factor—were identified as key experimental inputs. These variables were systematically altered at specific levels to investigate their impact on average recast layer thickness (AVLT), which serves as a crucial measure of surface integrity in micro-EDM machining. Beyond these process parameters, the influence of pulse generator type was also investigated to better understand its role in shaping the recast layer. Two distinct generators were assessed: the transistor pulse generator (TPG) and the IPG. The TPG, although commonly employed, delivers energy in an inconsistent and erratic manner due to its random pulse characteristics, which often result in irregular crater formation and a thicker, nonuniform recast layer. On the other hand, the IPG is a modification to provide uniform and stable energy discharges, leading to more controlled material erosion and enhanced surface quality.

4.2 Experimental Details

4.2.1 Materials and Method

The experiments were performed on a die-sinking EDM machine adapted for micro-EDM operations. The workpiece was mounted on a manually adjustable worktable submerged in a dielectric medium. The table position was fine-tuned using a conventional hand-wheel sliding mechanism. A tungsten carbide electrode (4 mm diameter) was fixed in the tool holder, which was connected to a servo motor and encoder system through a lead screw assembly to facilitate precise vertical tool movement. Controlled direct current (DC) pulses were supplied by a pulse generator connected between the tool and the workpiece. Two types of pulse generators were used: (i) a conventional transistor-based generator and (ii) a modified ISO energy pulse generator. A flushing system, integrated with the setup, ensured effective deionization of the dielectric fluid and removal of eroded particles from the spark gap. The contaminated dielectric fluid (EDM3 oil) was filtered and recirculated, with the flushing pressure maintained at 2 bar and adjustable via a dedicated control knob.

The workpiece material used in this study was Inconel 718, a nickel-based superalloy commonly used in aerospace applications due to its excellent mechanical strength, corrosion resistance, and high-temperature stability. Tungsten carbide was selected as the tool electrode material owing to its high melting point and wear resistance, which are critical for consistent micro-EDM performance.

4.2.2 Process Parameters

Three key electrical parameters—open circuit voltage, peak current, and duty factor—were selected as input variables for the micro-EDM experiments. These parameters are known to significantly influence the energy delivered during each discharge and, consequently, the quality of the machined surface [25]. Blind micro-holes were drilled to a depth of 2 mm on Inconel 718 samples under different parameter combinations and pulse generator conditions to assess their effect on surface integrity, particularly the recast layer formation. Table 4.1 summarizes the complete set of input variables used in the study.

The selection of an appropriate orthogonal array is a critical aspect of experimental design in the Taguchi method, which is employed in this study. The size of the chosen orthogonal array must be equal to or greater than the total degrees of freedom (DOF) of the experimental setup. The total DOF, including one degree of freedom allocated to the grand mean, can be calculated using Eq. 4.1.

$$\text{DOF} = F(P-1) + Q(P-1)^2 + 1 \tag{4.1}$$

Table 4.1 Process parameters in the micro-EDM process

Process parameters	Description
Workpiece	Inconel 718
Tool	Tungsten carbide
Dielectric medium	Total EDM3 oil
Flushing	Normally submerged
Drilling depth	2 mm
Pulse generator type	• Transistor pulse generator (TPG) • ISO pulse generator (IPG)
Gap voltage	• 40 V • 60 V • 80 V
Peak current	• 9 A • 12 A • 15 A
Duty factor	• 0.4 • 0.6 • 0.8

where P is the number of levels, F is the number of input factors, and Q is the number of considered interaction terms. In this study, three input factors are used, each with three levels, and interactions between factors are assumed to be negligible ($Q = 0$). Accordingly, the total DOF is 7, and an L9 orthogonal array, which accommodates up to 8 DOF, has been selected to efficiently conduct the micro-EDM experiments. Table 4.2 presents the L9 orthogonal array employed in this study.

Table 4.2 The L9 orthogonal array

Test	Gap voltage (V)	Peak current (A)	Duty factor
1	40	9	0.4
2	40	12	0.6
3	40	15	0.8
4	60	9	0.6
5	60	12	0.8
6	60	15	0.4
7	80	9	0.8
8	80	12	0.4
9	80	15	0.6

4.2.3 Evaluation of Recast Layer

During the micro-EDM process, both the workpiece and electrode surfaces experience localized melting. A portion of the molten material is expelled from the spark gap, while the remainder solidifies on the workpiece surface, forming a white layer commonly referred to as the recast layer. Due to limited impulse force in micro-EDM, complete evacuation of molten material is often not achieved, increasing the likelihood of recast layer formation. To evaluate surface integrity, the AVLT was measured using Scanning Electron Microscope (SEM) imaging. Owing to the inherent nonuniformity of the layer, multiple cross-sectional SEM images were analyzed to determine the average thickness. The recast layer boundary was digitally traced using AutoCAD 2013, where its area was computed and divided by the measurement length to yield thickness in micrometers (μm). This systematic analysis allowed for a quantitative comparison of the effects of electrical parameters and pulse generator types on the surface modification induced by the micro-EDM process.

4.3 Results and Discussion

After completing the experiments based on the L9 orthogonal array, the AVLT was measured for each test. In the micro-EDM process using the TPG, AVLT values ranged from 1.49 to 4.48 μm, indicating a relatively wide variation in thermal energy input and material redeposition. In contrast, the IPG exhibited a slightly narrower AVLT range, from 1.32 to 4.21 μm, suggesting more consistent energy distribution during discharge. To better interpret these results, main effects plots are presented in Fig. 4.1 and Fig. 4.2 for TPG and IPG-based processes, respectively. In the TPG-based process, both peak current and duty factor demonstrate significant and systematic effects on AVLT. Specifically, AVLT tends to increase with higher peak currents and larger duty factors. Gap voltage exhibits a relatively stable influence at 40 V and 60 V, where AVLT values are 2.67 μm and 2.59 μm, respectively. However, a noticeable increase to 2.98 μm is observed at 80 V, indicating a threshold beyond which excessive melting and redeposition occur. These variations can be attributed to the thermal nature of EDM, where increased discharge energy promotes deeper craters and more resolidified material, resulting in thicker recast layers. For the IPG-based micro-EDM process, similar trends are observed, though with reduced variation in AVLT, reflecting the improved control over energy delivery offered by the IPG system. Increasing peak current leads to a decrease in AVLT, likely due to more effective material ejection and improved flushing, which limits the accumulation of resolidified debris. Duty factor, on the other hand, shows a direct correlation with AVLT, as longer pulse-on durations promoted greater thermal input per discharge, increasing the extent of melting and recast layer formation. Gap voltage has only a minor influence in this configuration, with AVLT varying from 2.38 to 2.67 μm across the range of 40–80 V. No consistent trend is noted, indicating that within the

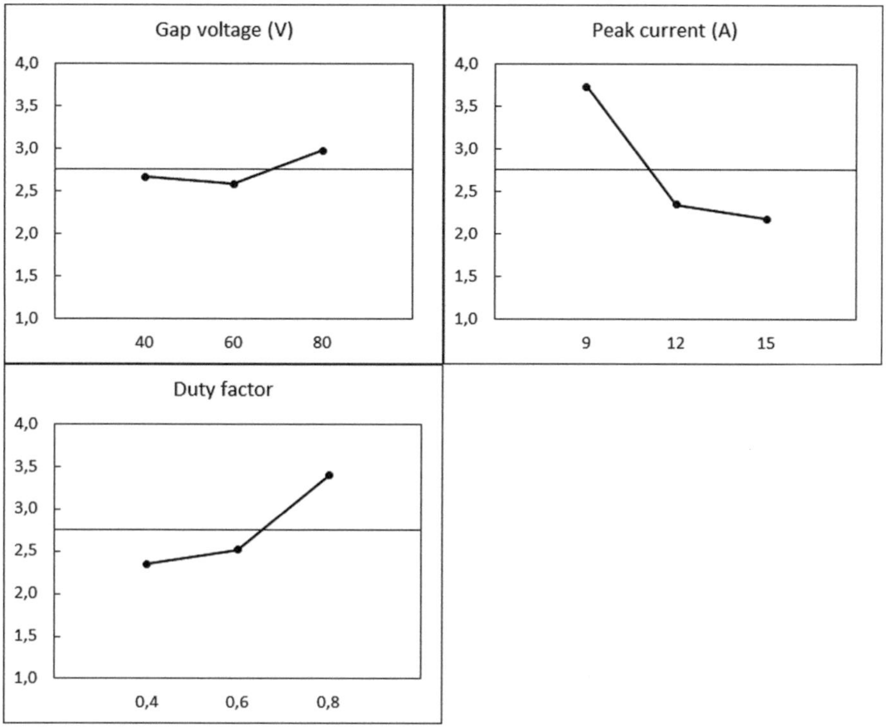

Fig. 4.1 Main effects plot for means with AVLT in TPG-based process

tested voltage levels, its effect is less dominant compared to peak current and duty factor.

From the overall AVLT measurements, it is evident that employing higher peak currents combined with lower duty factors is beneficial for minimizing recast layer thickness. Interestingly, the smallest AVLT for both pulse generator types is observed at a gap voltage of 60 V, despite the lack of a clearly linear relationship between voltage and AVLT. This may point to an optimal energy balance at this voltage level that minimizes redeposition without compromising machining efficiency. A comparison of pulse generator types reveals that the IPG-based micro-EDM process consistently yields better surface integrity. The grand mean of AVLT for the TPG-based process is 2.75 μm, whereas it decreases to 2.51 μm for the IPG-based configuration. This reduction highlights the role of refined energy control in reducing thermal damage and improving surface finish. Overall, the adoption of the IPG in micro-EDM applications appears to enhance machining performance by promoting uniform energy discharge, thereby reducing recast layer formation and associated surface defects.

Figure 4.3 presents the SEM images of recast layer formation for both TPG- and IPG-based micro-EDM processes. The TPG, due to its high-frequency operation and inherently random energy distribution, results in a nonuniform and discontinuous

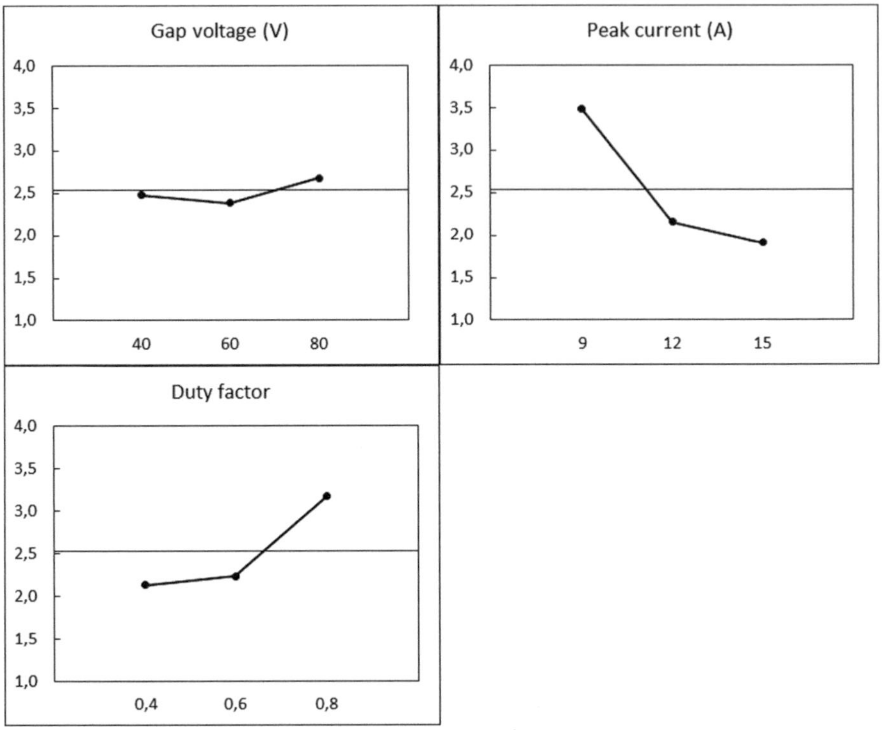

Fig. 4.2 Main effects plot for means with AVLT in IPG-based process

recast layer on the machined surface. This randomness leads to inconsistent thermal input, promoting irregular melting and redeposition of material across the surface. In contrast, the IPG delivers energy in a more controlled and uniformly distributed manner. Although it generates high temperatures sufficient for melting the material, the energy is released in a consistent and stable pulse pattern. As a result, the recast layer formed under IPG conditions is thinner, more uniform, and continuous in nature. Experimental observations confirm that the IPG significantly reduces recast layer formation compared to the TPG, regardless of the specific values of peak current, open circuit voltage, or duty factor [24]. This supports the conclusion that uniform energy delivery is a key factor in minimizing thermal damage and improving surface integrity in micro-EDM processes.

Figure 4.4 provides a schematic illustration of energy distribution over the micro-machined surface using different pulse generator types. In the case of the TPG, the nonlinear nature of its pulse train leads to highly variable energy discharges. This causes larger and irregular crater formation, as well as increased surface roughness due to inconsistent spark erosion. On the other hand, the IPG offers improved control by delivering pulses with equal and consistent energy levels, especially during the finishing stages of machining. This uniform discharge not only reduces crater size variation but also contributes to smoother surface profiles and more stable machining

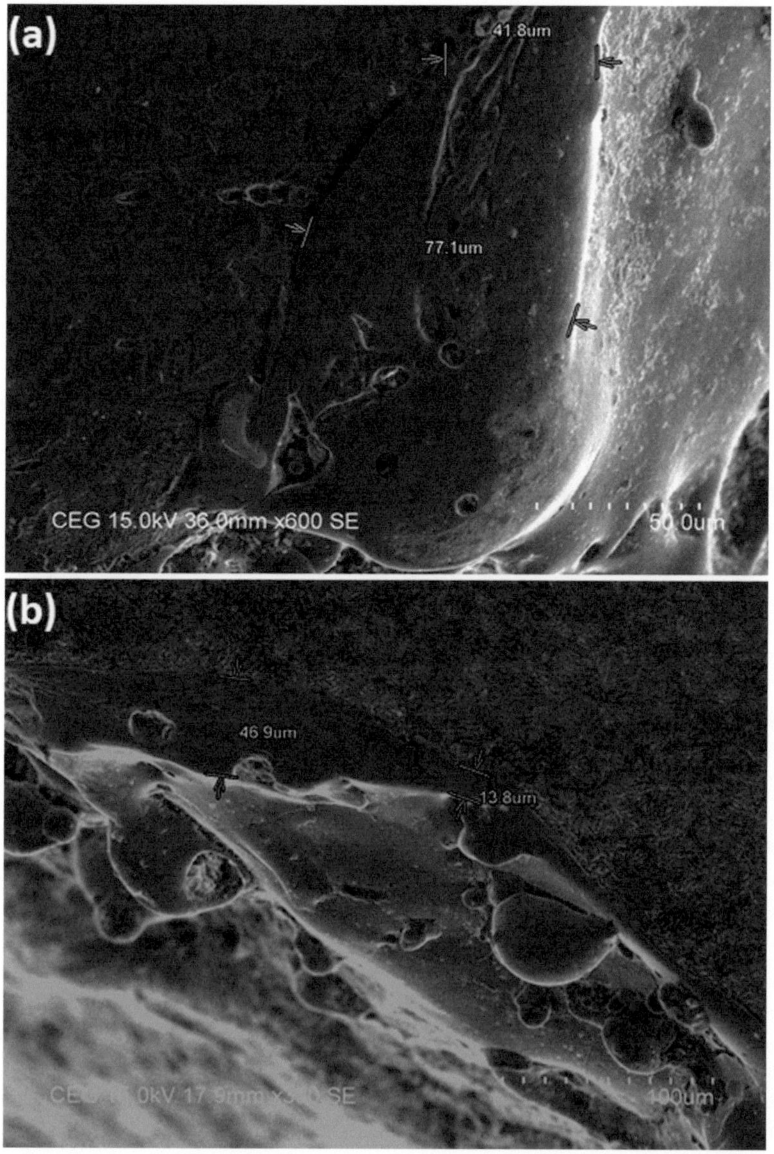

Fig. 4.3 SEM image of recast layer formations in **a** TPG- and **b** IPG-based micro-EDM process

conditions. The comparative analysis of SEM images and energy distribution patterns reveals that the IPG provides a distinct advantage over the TPG by producing more consistent discharge energy. This enhancement results in reduced recast layer thickness, lower surface roughness, and improved overall surface quality, making the

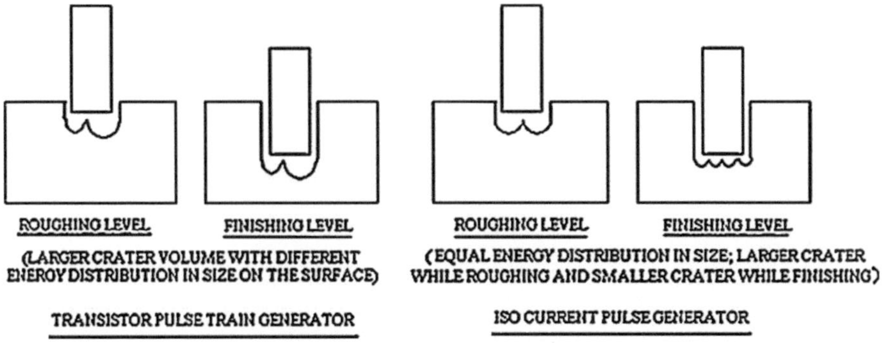

Fig. 4.4 Energy distribution on the surface with different pulse generators [24]

IPG a more suitable choice for precision micro-EDM applications, especially when machining difficult-to-cut aerospace alloys such as Inconel 718.

Similar results were reported by Muthuramalingam and Mohan [26], who also observed that the IPG methodology leads to reduced recast layer formation and improved surface quality compared to the TPG approach, aligning closely with the findings of the current study. On average, the IPG achieves a 30.4% overall reduction in recast layer thickness relative to the TPG. When using a copper tool electrode, the IPG reduces the recast layer by 38.7%, whereas with a brass tool electrode, the reduction is 23.1%. Similarly, with a tungsten carbide tool electrode, the IPG leads to a 29.6% decrease in recast layer thickness compared to the TPG. The thickness of the recast layer is primarily influenced by process parameters such as gap voltage, peak current, and duty factor. Since voltage and current are directly related in the micro-EDM process, the recast layer thickness is strongly governed by the amount of thermal energy generated during discharge, which is proportional to the temperature reached at the spark site on both the tool and the workpiece surfaces.

It is also important to note that the effect of electrode material is also significant on the recast layer thickness. Copper has a melting point of 1085 °C, which is higher than that of brass (940 °C), while tungsten carbide has the highest melting point among the three at 2870 °C. These variations in melting point directly influence the material erosion, melting behavior, and consequently, the extent of recast layer formation during the EDM process. For instance, higher melting point materials like tungsten carbide tend to experience less tool wear and localized melting, thereby resulting in thinner and more stable recast layers. On the other hand, lower melting point electrodes may contribute more molten material during discharges, leading to increased redeposition and thicker recast layers. In addition to pulse generator type, tool electrode material plays critical roles in determining the thermal behavior at the machining interface and the resultant surface integrity. The superior performance of the IPG in reducing recast layer formation across all electrode types reinforces its suitability for precision applications involving high-performance materials such as Inconel 718 [24, 26].

4.4 Conclusions

In this study, the micro-EDM process was applied to Inconel 718, a high-performance nickel-based superalloy known for its excellent mechanical strength, corrosion resistance, and thermal stability. Due to these properties, Inconel 718 is widely used in critical aerospace applications such as turbine blades, engine components, and exhaust systems. However, its high toughness and low thermal conductivity make it extremely challenging to machine using conventional methods. As a result, micro-EDM presents itself as a promising nontraditional machining technique capable of producing precise features with minimal mechanical stresses and tool wear.

To optimize the process, three critical electrical parameters—open circuit voltage, peak current, and duty factor—were selected as the primary control variables in the experimental design. These parameters were varied across defined levels to investigate their influence on the AVLT, a key indicator of surface integrity in micro-EDM operations. The recast layer, formed due to the rapid melting and solidification of material at the discharge site, can significantly affect the mechanical properties and fatigue life of machined components if not properly controlled. In addition to these electrical parameters, the type of pulse generator was also investigated to assess its contribution to the formation and morphology of the recast layer. Two types of generators, TPG and IPG, were evaluated. The TPG, while widely used, is known for its irregular and nonuniform energy delivery due to random pulse characteristics, often leading to uneven crater formation and thicker, discontinuous recast layers. In contrast, the IPG is designed to deliver energy in a more consistent and controlled manner, thereby enabling more uniform thermal erosion and improved surface characteristics.

Experimental results demonstrated that gap voltage had a relatively minor effect on AVLT, with small variations observed across the voltage range of 40–80 V in both TPG and IPG modes. However, peak current and duty factor exhibited pronounced effects. An increase in peak current led to a reduction in AVLT, likely due to the generation of higher-intensity discharges that enhance material removal efficiency while reducing molten material redeposition. Conversely, higher duty factors were associated with increased AVLT, as prolonged pulse duration leads to greater heat accumulation and a higher likelihood of molten material resolidifying on the surface. Beyond the individual parameter effects, the pulse generator type itself was a decisive factor in determining machining performance. The IPG consistently produced thinner, more uniform recast layers with fewer microcracks and surface irregularities compared to the TPG. This improvement can be attributed to the IPG ability to maintain stable discharge energy, avoid arcing, and facilitate efficient flushing of molten debris during machining. Moreover, the more uniform crater geometry and reduced thermal damage observed under IPG conditions are advantageous for applications where dimensional accuracy and surface finish are critical. These findings exhibit the importance of both precise control of electrical parameters and pulse generator technology in achieving superior micro-EDM performance, particularly for difficult-to-machine aerospace alloys like Inconel 718. The adoption of the IPG in industrial micromachining setups could lead to better component reliability, longer fatigue life, and reduced need for postprocessing treatments.

References

1. K. H. Ho and S. T. Newman, "State of the art electrical discharge machining (EDM)," *Int. J. Mach. Tools Manuf.*, vol. 43, no. 13, pp. 1287–1300, Oct. 2003, https://doi.org/10.1016/S0890-6955(03)00162-7.
2. D. Sharma and S. S. Hiremath, "Review on tools and tool wear in EDM," *Mach. Sci. Technol.*, vol. 25, no. 5, pp. 802–873, Sep. 2021, https://doi.org/10.1080/10910344.2021.1971711.
3. M. Bellotti, M. Wu, J. Qian, and D. Reynaerts, "Tool Wear and Material Removal Predictions in Micro-EDM Drilling: Advantages of Data-Driven Approaches," *Appl. Sci.*, vol. 10, no. 18, p. 6357, Sep. 2020, https://doi.org/10.3390/app10186357.
4. B. Mohan, A. Rajadurai, and K. G. Satyanarayana, "Effect of SiC and rotation of electrode on electric discharge machining of Al–SiC composite," *J. Mater. Process. Technol.*, vol. 124, no. 3, pp. 297–304, Jun. 2002, https://doi.org/10.1016/S0924-0136(02)00202-9.
5. M.-D. Moses and M. P. Jahan, "Micro-EDM machinability of difficult-to-cut Ti-6Al-4V against soft brass," *Int. J. Adv. Manuf. Technol.*, vol. 81, no. 5–8, pp. 1345–1361, Nov. 2015, https://doi.org/10.1007/s00170-015-7306-9.
6. M. P. Jahan, Y. S. Wong, and M. Rahman, "A study on the quality micro-hole machining of tungsten carbide by micro-EDM process using transistor and RC-type pulse generator," *J. Mater. Process. Technol.*, vol. 209, no. 4, pp. 1706–1716, Feb. 2009, https://doi.org/10.1016/j.jmatprotec.2008.04.029.
7. J. Wang, Q. Gao, X.-C. Xi, and W.-S. Zhao, "Experimental study of EDM milling of 3D-shaped diffuser for film cooling holes on turbine blades," *Procedia CIRP*, vol. 113, pp. 160–165, 2022, https://doi.org/10.1016/j.procir.2022.09.125.
8. H. Tong, Y. Li, L. Zhang, and B. Li, "Mechanism design and process control of micro EDM for drilling spray holes of diesel injector nozzles," *Precis. Eng.*, vol. 37, no. 1, pp. 213–221, Jan. 2013, https://doi.org/10.1016/j.precisioneng.2012.09.004.
9. L. Li, C. Diver, J. Atkinson, R. Giedl-Wagner, and H. J. Helml, "Sequential Laser and EDM Micro-drilling for Next Generation Fuel Injection Nozzle Manufacture," *CIRP Ann.*, vol. 55, no. 1, pp. 179–182, 2006, https://doi.org/10.1016/S0007-8506(07)60393-X.
10. K. Janardhana, N. K. Anushkannan, K. P. Dinakaran, R. K. Puse, and S. Boopathi, "Experimental investigation on microhardness, surface roughness, and white layer thickness of dry EDM," *Eng. Res. Express*, vol. 5, no. 2, p. 025022, Jun. 2023, https://doi.org/10.1088/2631-8695/acce8f.
11. H. Sidhom, F. Ghanem, T. Amadou, G. Gonzalez, and C. Braham, "Effect of electro discharge machining (EDM) on the AISI316L SS white layer microstructure and corrosion resistance," *Int. J. Adv. Manuf. Technol.*, vol. 65, no. 1–4, pp. 141–153, Mar. 2013, https://doi.org/10.1007/s00170-012-4156-6.
12. S. Caruso, D. Umbrello, J. C. Outeiro, L. Filice, and F. Micari, "An Experimental Investigation of Residual Stresses in Hard Machining of AISI 52100 Steel," *Procedia Eng.*, vol. 19, pp. 67–72, 2011, https://doi.org/10.1016/j.proeng.2011.11.081.
13. N. S. Rossini, M. Dassisti, K. Y. Benyounis, and A. G. Olabi, "Methods of measuring residual stresses in components," *Mater. Des.*, vol. 35, pp. 572–588, Mar. 2012, https://doi.org/10.1016/j.matdes.2011.08.022.
14. H. T. Lee and T. Y. Tai, "Relationship between EDM parameters and surface crack formation," *J. Mater. Process. Technol.*, vol. 142, no. 3, pp. 676–683, Dec. 2003, https://doi.org/10.1016/S0924-0136(03)00688-5.
15. B. Ekmekci, "Residual stresses and white layer in electric discharge machining (EDM)," *Appl. Surf. Sci.*, vol. 253, no. 23, pp. 9234–9240, Sep. 2007, https://doi.org/10.1016/j.apsusc.2007.05.078.

16. C. A. Huang, F. Y. Hsu, and S. J. Yao, "Microstructure analysis of the martensitic stainless steel surface fine-cut by the wire electrode discharge machining (WEDM)," *Mater. Sci. Eng. A*, vol. 371, no. 1–2, pp. 119–126, Apr. 2004, https://doi.org/10.1016/j.msea.2003.10.277.
17. H.-T. Lee, F.-C. Hsu, and T.-Y. Tai, "Study of surface integrity using the small area EDM process with a copper–tungsten electrode," *Mater. Sci. Eng. A*, vol. 364, no. 1–2, pp. 346–356, Jan. 2004, https://doi.org/10.1016/j.msea.2003.08.046.
18. N. Mohd Abbas, D. G. Solomon, and Md. Fuad Bahari, "A review on current research trends in electrical discharge machining (EDM)," *Int. J. Mach. Tools Manuf.*, vol. 47, no. 7–8, pp. 1214–1228, Jun. 2007, https://doi.org/10.1016/j.ijmachtools.2006.08.026.
19. M.-C. Shin, Y.-S. Kim, H.-G. Cheong, and C.-N. Chu, "Performance of a TR-iso-pulse generator in micro ED-drilling," *Precis. Eng.*, vol. 56, pp. 438–446, Mar. 2019, https://doi.org/10.1016/j.precisioneng.2019.01.013.
20. H. Mishra, Y. K. Modi, S. K. Tiwari, and K. Maity, "Experimental investigation of micro-EDM hole drilling on Inconel-718 for predicting surface quality characteristics," *J. Micromanufacturing*, p. 25165984251328100, May 2025, https://doi.org/10.1177/25165984251328100.
21. T. Kawakami and M. Kunieda, "Study on Factors Determining Limits of Minimum Machinable Size in Micro EDM," *CIRP Ann.*, vol. 54, no. 1, pp. 167–170, 2005, https://doi.org/10.1016/S0007-8506(07)60075-4.
22. A. P. Tiwary, B. B. Pradhan, and B. Bhattacharyya, "Study on the influence of micro-EDM process parameters during machining of Ti–6Al–4V superalloy," *Int. J. Adv. Manuf. Technol.*, vol. 76, no. 1–4, pp. 151–160, Jan. 2015, https://doi.org/10.1007/s00170-013-5557-x.
23. C. P. Mohanty, J. Sahu, and S. S. Mahapatra, "Thermal-structural Analysis of Electrical Discharge Machining Process," *Procedia Eng.*, vol. 51, pp. 508–513, 2013, https://doi.org/10.1016/j.proeng.2013.01.072.
24. T. Muthuramalingam and B. Mohan, "Performance analysis of iso current pulse generator on machining characteristics in EDM process," *Arch. Civ. Mech. Eng.*, vol. 14, no. 3, pp. 383–390, May 2014, https://doi.org/10.1016/j.acme.2013.10.003.
25. B. Mohan, A. Rajadurai, and K. G. Satyanarayana, "Electric discharge machining of Al–SiC metal matrix composites using rotary tube electrode," *J. Mater. Process. Technol.*, vol. 153–154, pp. 978–985, Nov. 2004, https://doi.org/10.1016/j.jmatprotec.2004.04.347.
26. T. Muthuramalingam and B. Mohan, "Influence of Tool Electrode Properties on Machinability in Spark Erosion Machining," *Mater. Manuf. Process.*, p. 130122112650006, Jan. 2013, https://doi.org/10.1080/10426914.2013.763973.

Chapter 5
Near-Net Shape Microholes Formed by Micropunch

Kelvii Wei Guo

Abstract Microhole fabrication via micropunching is a challenging operation within the framework of micromanufacturing. The dimensional accuracy of the fabricated microholes is directly affected by wear formation on the micropunch surface. Therefore, the punching shot count must be carefully determined during process design, taking into account the wear-induced degradation of the tool. In conventional micropunching operations, hard materials such as tungsten carbide and cobalt (WC/Co) based materials are commonly used to manufacture micropunches. However, even with such hard materials, tool wear significantly limits the number of effective punching shots. In this study, near-net shape microholes were fabricated in commercially pure titanium workpieces using WC/Co micropunches, and the wear progression of the micropunch was monitored with respect to punching shot count. The results showed that the WC/Co micropunches maintained acceptable wear levels and dimensional accuracy up to approximately 1250 shots. To further extend the service life of the micropunches, multiwalled carbon nanotube (MWCNT) coatings were applied to the working region of the WC/Co micropunches. The micropunching process was then repeated under the same conditions. According to the experimental results, the MWCNT coating effectively extended the service life, increasing the acceptable punching shot count to approximately 1425 shots, while maintaining dimensional accuracy and minimizing wear loss. This improvement is attributed to the self-lubricating properties of the MWCNTs, which formed a protective layer or transfer film at the contact interface, owing to their inherent graphitic structure.

Keywords Near-net shape manufacturing · Microhole formation · Micropunch · Wear · Multi-walled carbon nanotubes

---

K. W. Guo (✉)
Department of Mechanical and Biomedical Engineering, City University of Hong Kong, Kowloon Tong, Kowloon, Hong Kong
e-mail: kelviiguo@yahoo.com

S. Gürgen (ed.), *Micromanufacturing Operations*,
https://doi.org/10.1007/978-3-032-04193-7_5

5.1 Introduction

Challenges in microscale manufacturing are rapidly increasing due to the significant advancements in miniaturization and microfabrication, particularly in fields such as medical implants, telecommunications, microscale pumps, microchemical reactors, and microscale batteries [1–4]. In line with current green and eco-friendly requirements, it is crucial to fabricate microholes on a large scale with high efficiency [5–7]. Today, both industry and daily life demand small-sized products that are low-cost, high-quality, and mass-producible. Therefore, there is an urgent need to develop techniques that enable large-scale production while ensuring low cost, environmental sustainability, and high precision [8, 9].

Manufacturing processes can be classified based on the type of energy utilized, including electrical, chemical, electrochemical, mechanical, and laser-based methods. These processes inevitably involve variations such as thermal effects, mechanical forces, dissolution, solidification, and recomposition. Furthermore, manufacturing can generally be categorized into additive, subtractive, forming, joining, and hybrid methods, a classification that also applies to microfabrication. Additive processes encompass techniques such as chemical deposition, direct writing, microinjection molding, microcasting, and coating methods like physical vapor deposition (PVD), chemical vapor deposition (CVD), and photoelectron forming. In contrast, subtractive processes include methods such as microelectrochemical machining (micro-ECM), micro-electrical discharge machining (micro-EDM), micromechanical cutting (e.g., milling, turning, grinding, polishing), photochemical machining, electron beam machining, and laser beam machining. Forming processes include microimprinting/nanoimprinting, hot embossing, and microforming techniques such as superplastic forming, extrusion, deep drawing, forging, bending, stamping, punching, and hydroforming. Joining processes, on the other hand, involve methods like gluing, bonding/welding, soldering, resistance welding, micromechanical assembly, and laser joining. Beyond these techniques, hybrid processes include LIGA (Lithographie, Galvanoformung, and Abformung), LIGA combined with laser machining or micro-EDM, and micro-laser-ECM [10–17].

Punching is a feasible and straightforward method for producing holes on a large scale. It is popular, effective, and efficient. Compared to the processes mentioned above, punching offers several notable advantages, including lower costs, reduced waste, better material integrity, and higher production rates, making it particularly suitable for tailoring microproducts in mass fabrication. As a result, micropunching is considered an optimal approach for fabricating large quantities of microproducts, especially when considering the clearance between the microdie and micropunch. Furthermore, it is environmentally friendly, as it does not involve any chemicals [18]. Therefore, as an eco-friendly and cost-effective method for producing microholes on a large scale, micropunching is highly effective [19–27].

Due to the interaction between the micropunch and the workpiece, tool wear (specifically, wear of the micropunch) is a critical issue in micropunching. Even when punches are made from tough and hard materials, such as cobalt (Co) and tungsten carbide (WC), the quality of the fabricated holes rapidly deteriorates under the repeated effects of successive punching. Therefore, to provide a more promising approach for microfabrication, this study investigates the variation in microhole quality at different punching stages using a WC/Co micropunch.

5.2 Experimental Details

Figure 5.1 shows the details of the micropunch with a diameter of 150 μm, which was fabricated by an ultrahard tools company in Japan. The micropunch is composed of Co particles (25% volume fraction) and WC particles (75% volume fraction) with a mean particle size of 50 μm. The workpiece used is a pure titanium (Ti) sheet with a thickness of 200 μm.

The distribution of elements in the fabricated micropunch and the corresponding surface microtexture are shown in Fig. 5.2. As shown, ploughs induced by fabrication are clearly scattered across the ultra-precisely machined surface, along with the presence of some debris. Moreover, the density of the tungsten (W) element on the micropunch surface is higher than that of the carbon (C) and cobalt (Co) elements, indicating that the fabricated micropunch, with its hard surface, is an ideal candidate for micropunching.

To obtain the required experimental surfaces, pure titanium sheets were carefully cleaned with ethyl alcohol and acetone. After cleaning, the workpieces were processed using an MP50 microprocessing machine (made in Japan) with parameters of 20 pulses per minute, a feed rate of 2 mm, and a load of 15 N. The experiments were conducted using a homemade setup, as depicted in Fig. 5.3. To ensure the wear resistance and durability of the punching tool, as well as the quality of the microholes, vertical alignment is crucial for achieving better punching quality. Therefore, the fixture was precisely controlled with a vertical motion carriage that had a lateral error of less than 1.5 μm over a maximum vertical stroke of 20 mm. A push rod, driven by a Z-stepper, applied the stroke and load (up to 20 N) to the carriage within an accurate digital control system. Moreover, the bush was fabricated from WC and

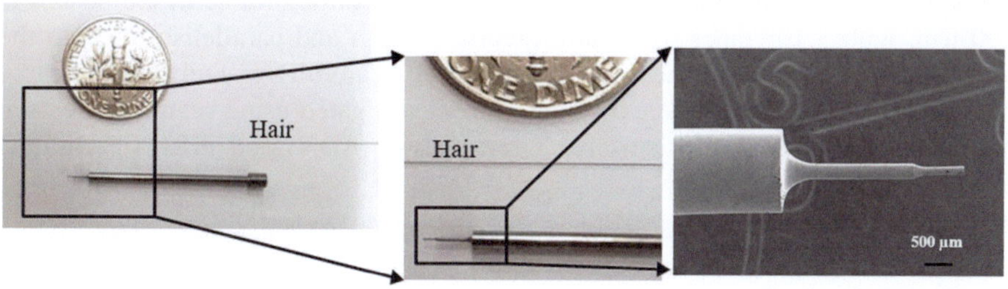

Fig. 5.1 Detailed views of the micropunch

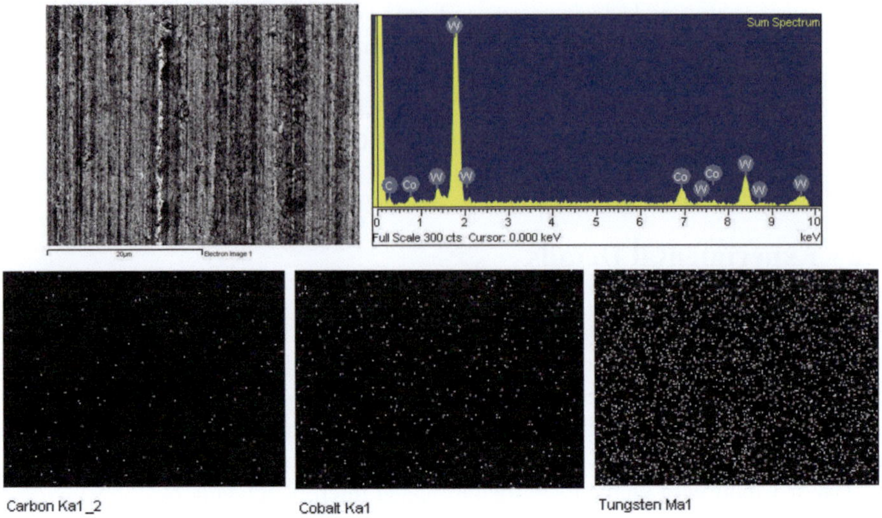

Fig. 5.2 Elemental distribution of the micropunch

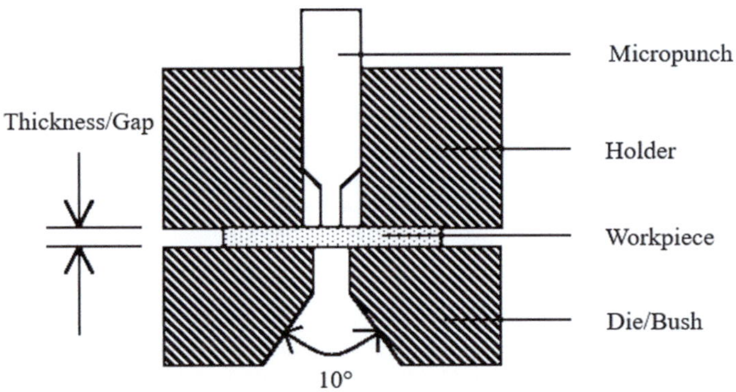

Fig. 5.3 Schematic illustration of the micropunching process

Co, similar to the micropunch. The diameter of the through-hole in the bushes is 200 μm, with a 10° taper. To ensure precise flatness and parallelism of both the upper and lower faces, the faces of the bushes were carefully ground. The details of microhole alterations at different stages were investigated using scanning electron microscopy (SEM), energy dispersive X-ray analysis (EDX), and an LEXT confocal laser-OLS3000.

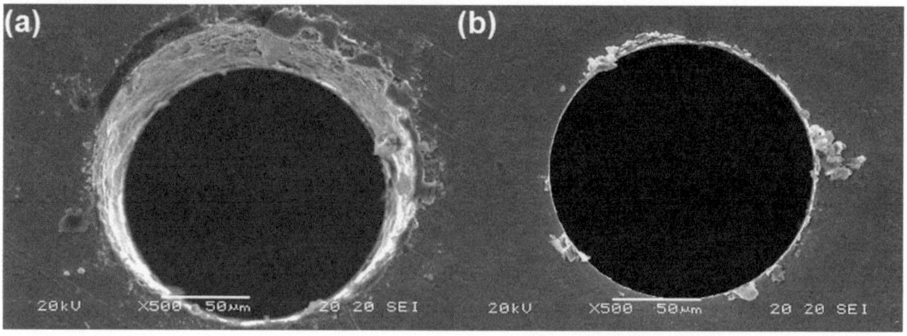

Fig. 5.4 Microhole morphology in the initial stage of micropunching: **a** front side and **b** back side

5.3 Results and Discussion

5.3.1 Effect of Punching Shots (<500) on the Morphology of Punched Microholes

Figure 5.4 shows the morphology of the punched microhole after fewer than 500 punching shots, indicating the initial stage of micropunching. It illustrates the presence of debris on both sides of the microhole, particularly on the front side. Figure 5.5 shows the corresponding EDX results, which reveal that the debris primarily originates from the workpiece (Ti) and the micropunch (C, Co, W). In addition to these base elements, due to contamination during the micropunching process, elements such as sodium (Na), chlorine (Cl), sulfur (S), and oxygen (O) are also detected. The debris on the backside is less concentrated than on the front side. Clearly, the punched microhole at this stage is not suitable for direct practical applications.

Figure 5.6 depicts the microtexture of the micropunch after fewer than 500 punching shots. Compared to Fig. 5.2, it is noteworthy that the wear on the micropunch has changed significantly, with the ploughs becoming flatter due to the interaction between the micropunch and the workpiece. The initial plough tracks have almost disappeared, remaining only in scattered areas. Moreover, hard particles, such as WC, are no longer visible on the micropunch surface. This indicates that, in the initial stage, the wear of the micropunch is primarily due to the loss of Co and the removal of the original debris.

5.3.2 Effect of Punching Shots (500–1200) on the Morphology of Punched Microholes

Figure 5.7 shows the effect of punching shots (500–1200) on the morphology of the punched microhole. The results show that while some workpiece debris is clearly

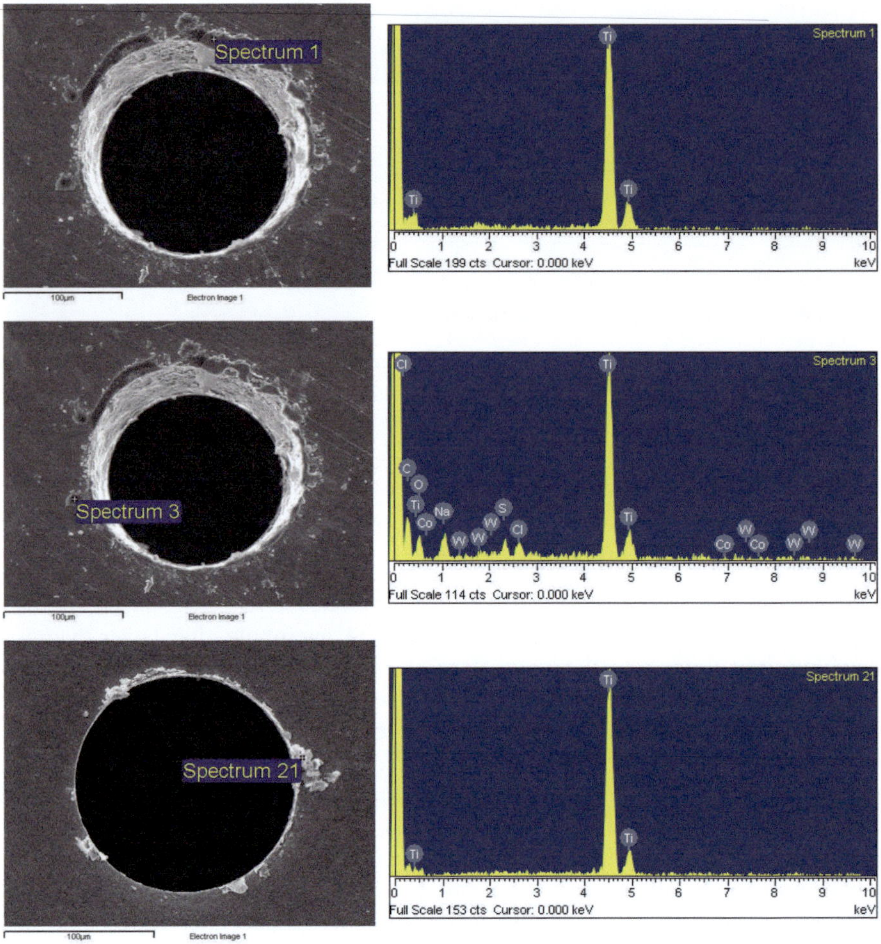

Fig. 5.5 EDX analysis of the debris

distributed on the backside, the overall profile of the punched microhole remains well-defined and within the desired dimensions. Compared to Fig. 5.4, the quality of punched microholes in this period (quasi-stable period) is significantly improved compared to the initial stage. This suggests that due to the combined effect of WC and Co in the micropunch, wear loss remains minimal during the 500–1200 punching shot range. Consequently, the punched microholes maintain high quality. Moreover, as the number of punching shots increases, the wear loss of the micropunch stabilizes. Compared to the initial stage, the wear in the 500–1200 shot counts is significantly lower, indicating that the micropunch is in a quasi-stable wear condition.

Figure 5.8 shows the surface microtexture of the micropunch during the quasi-stable period. It reveals that WC particles are clearly visible on the micropunch

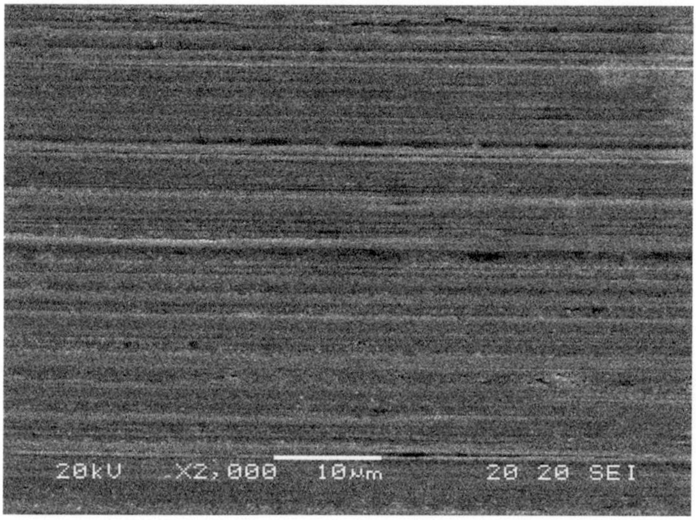

Fig. 5.6 Micropunch surface microtexture after fewer than 500 punching shots

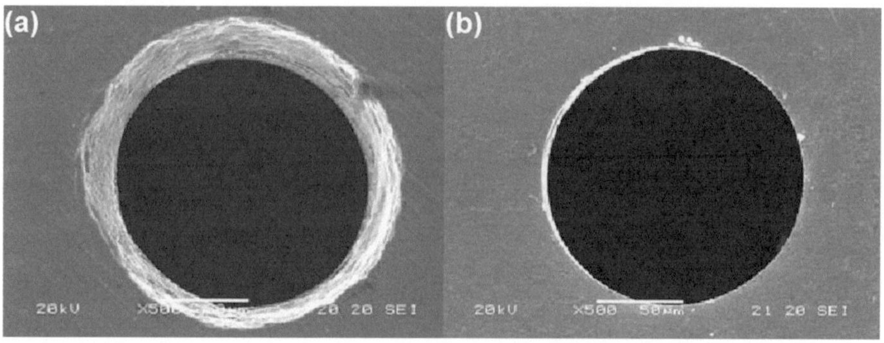

Fig. 5.7 Microhole morphology in the quasi-stable period: **a** front side and **b** back side

surface at this stage. Moreover, the uniform distribution of WC particles enhances the wear resistance of the micropunch during processing.

5.3.3 *Effect of Punching Shots (>1200) on the Morphology of Punched Microholes*

When the number of punching shots exceeds 1200, the morphology of the punched microhole is shown in Fig. 5.9. It can be observed that workpiece debris adheres to both sides of the microhole, especially on the inner surface. These punched

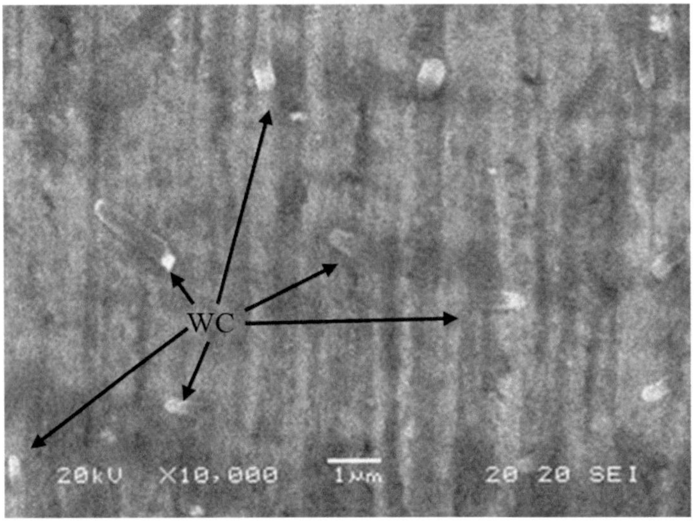

Fig. 5.8 Micropunch surface microtexture after 500–1200 punching shots

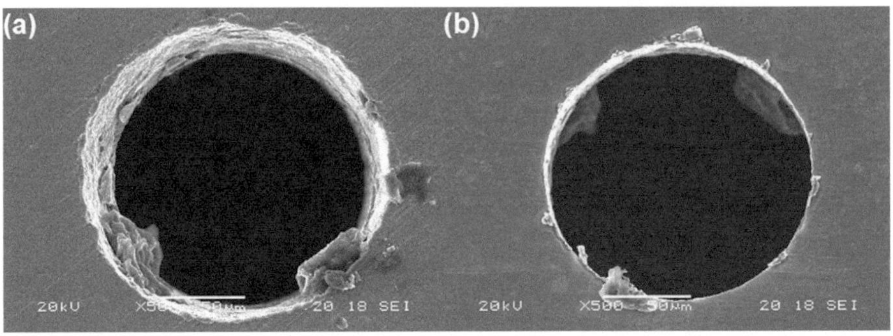

Fig. 5.9 Microhole morphology in the severe wear period: **a** front side and **b** back side

microholes cannot be directly utilized in practical applications, which limits their usability. With an increasing number of punching shots, the wear of the micropunch becomes severe. Consequently, micropunching becomes ineffective, and titanium from the workpiece adheres to the edges of the microhole due to the failure of material removal. Simultaneously, the titanium that remains stuck to the microhole edge is sheared under the continuous feed motion, leading to the formation of larger debris, as shown in Fig. 5.10. Furthermore, as the number of punching shots continues to increase, the wear condition of the micropunch enters an intensively severe stage. At this point, micropunch material is severely peeled off, while detached material strongly adheres to the titanium workpiece.

Figure 5.11 shows the surface microtexture of the micropunch during the severe wear period. Compared to Fig. 5.7, it is evident that a large number of WC particles

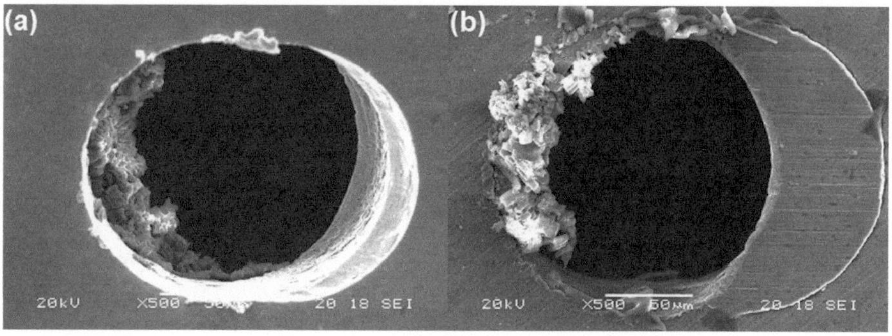

Fig. 5.10 Microhole morphology in the intensively severe wear period: **a** front side and **b** back side

Fig. 5.11 Micropunch surface microtexture after greater than 1200 punching shots

are distributed on the micropunch surface in this stage. Furthermore, due to the significant wear loss of Co—which serves as the bonding material in micropunch fabrication—maintaining precise alignment becomes challenging, ultimately leading to the failure of the micropunching process. This indicates that in the severe wear period, the primary characteristics of the micropunch are dominated by WC particles, as Co loss becomes extensive. Consequently, the synergistic effect between Co and WC disappears, causing WC particles to detach easily and adhere to the formed microholes.

5.3.4 Features of the Microholes

Figure 5.12 shows the diameter measurements of a punched microhole, obtained using the LEXT confocal laser OLS3000. The corresponding results, measured five times for both sides, are illustrated in Fig. 5.13. Accordingly, Fig. 5.14 shows the relative wear loss of the micropunch at different punching shots. Considering these results, it is evident that the morphology of the punched microholes evolves in correlation with the wear progression of the micropunch. Moreover, the variations across different periods are distinct, and the observed changes in microhole diameter align with the wear loss of the micropunch. It is clear that during the quasi-stable period, the wear loss of the micropunch remains nearly constant with an increase in punching shots (with near-zero weight loss), particularly for punching shots between 500 and 1200. As a result, the diameter of the punched microholes remains remarkably consistent.

For punching shots greater than 1525 (highlighted by the purple circle), the quality of the punched microholes significantly deteriorates, with a noticeable increase in micropunch wear loss and unsatisfactory profiles of the punched microholes. During the severe punching period, WC particles are easily peeled off due to the substantial loss of the bonding material, Co. Furthermore, considering the friction between the micropunch and the workpiece, the temperature of the micropunch increases significantly as the punching process progresses. Due to the different thermal expansion coefficients of Co and WC, the WC particles become more prone to detachment, increasing the wear loss of the micropunch. Consequently, more drastic wear loss occurs as the feed rate of the micropunching process increases.

It is also seen that, for punching shots ranging from 1250 to 1500, the wear loss of the micropunch changes dramatically, and the alteration of the micropunch becomes wavily unstable. Specifically, when punching shots exceed 1250, the wear loss of the micropunch increases significantly. During the period between 1300 and 1400

Fig. 5.12 Microhole diameter measurements

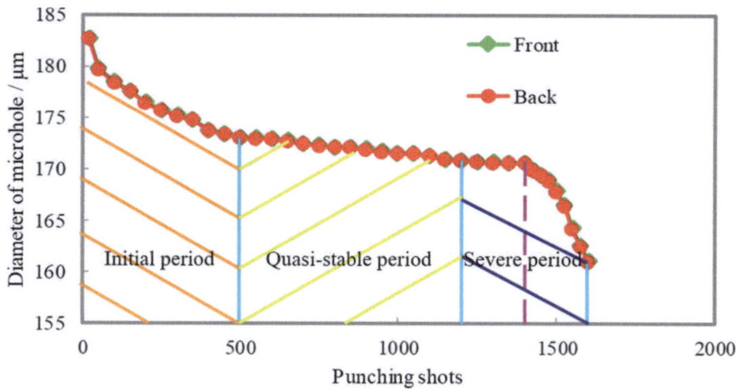

Fig. 5.13 Relationship between microhole diameter and punching shots

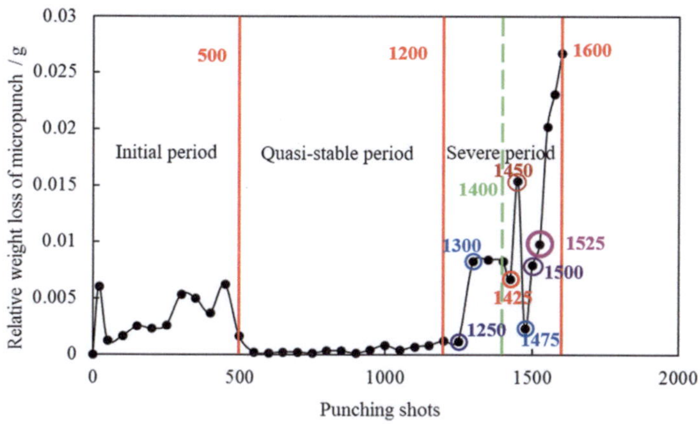

Fig. 5.14 Relationship between micropunch wear loss and punching shots

punching shots, the wear loss of the micropunch stabilizes, but at a level higher than that observed during the initial and quasi-stable periods; this is referred to as the pseudo-stable period. As punching shots continue (between 1400 and 1500), the wavy nature of the wear loss becomes more apparent, with the wear loss fluctuating severely. Notably, between 1400 and 1425 punching shots, the wear loss of the micropunch slightly decreases. However, when punching shots exceed 1425, the wear loss increases sharply, reaching its maximum value around 1450 shots, marking the midpoint of the severe period. After this peak, the wear loss decreases steeply, reaching its minimum at 1475 shots. As the punching shots increase beyond 1475, the wear loss gradually rises again, approaching a value similar to that of the pseudo-stable period at 1500 shots. This indicates that after 1525 punching shots, the wear loss of the micropunch steadily increases. Consequently, the micropunching process becomes entirely compromised, resulting in the failure of the punched microholes.

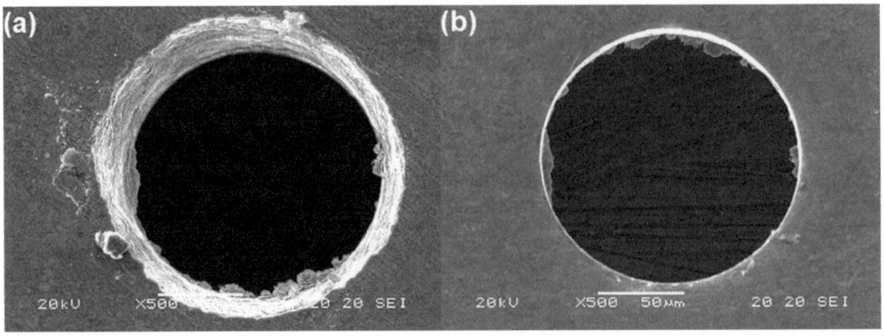

Fig. 5.15 Microhole morphology at the beginning of the severe period: **a** front side and **b** back side

Interestingly, during the so-called severe period (punching shots >1200), the diameter of the punched microhole remains relatively stable at the beginning of this period (i.e., punching shots between 1200 and 1400), as indicated by the light blue solid line and purple dashed line in Fig. 5.13. Correspondingly, Fig. 5.14 shows the variation in the micropunch during this period, as marked by the red solid line and green dashed line. It is observed that although the wear loss of the micropunch increases noticeably, it remains relatively stable. The corresponding profile of the punched microhole during this period is depicted in Fig. 5.15. The results show that debris is distributed both inside and outside the microhole, particularly on the front side (Fig. 5.15a). As a result, the quality of the punched microhole has deteriorated, even though the backside remains comparatively clean (Fig. 5.15b). Furthermore, it is noteworthy that even during the severe period, the diameter of the punched microhole on the front side is nearly the same as that on the back side.

Recalling Fig. 5.7, along with the microhole diameter and wear loss results, it can be stated that, without any post-treatment, the punched microholes in the quasi-stable period are of sufficient quality for direct use. This makes them highly suitable for near-net shape fabrication in micromanufacturing.

The distinguishing features of conventional punching are shown in Fig. 5.16, which includes four sections: (a) rollover, (b) shear/burnish zone, (c) fractured zone, and (d) burr. However, the features of micropunching with foil thicknesses less than 200 μm differ significantly. Figure 5.17 shows the key features of the punched microhole. Compared to conventional punching, the differences are evident. It can be seen that, for micropunching, the fracture band is absent. Moreover, the areas of rollover and burr are significantly narrower than those in conventional punching. Furthermore, the shear/burnish band constitutes the majority of the inner microhole.

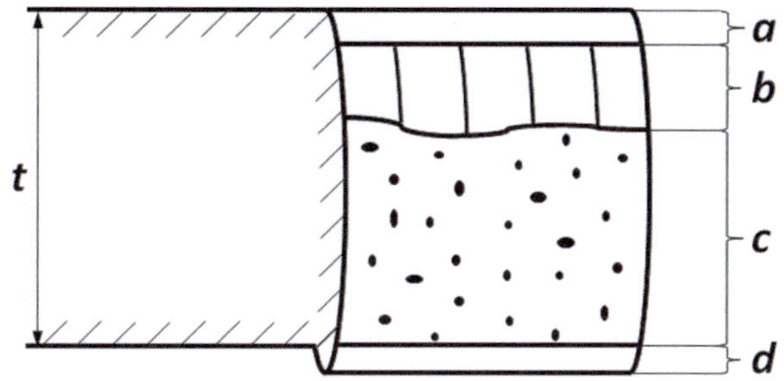

Fig. 5.16 Features of conventional punching

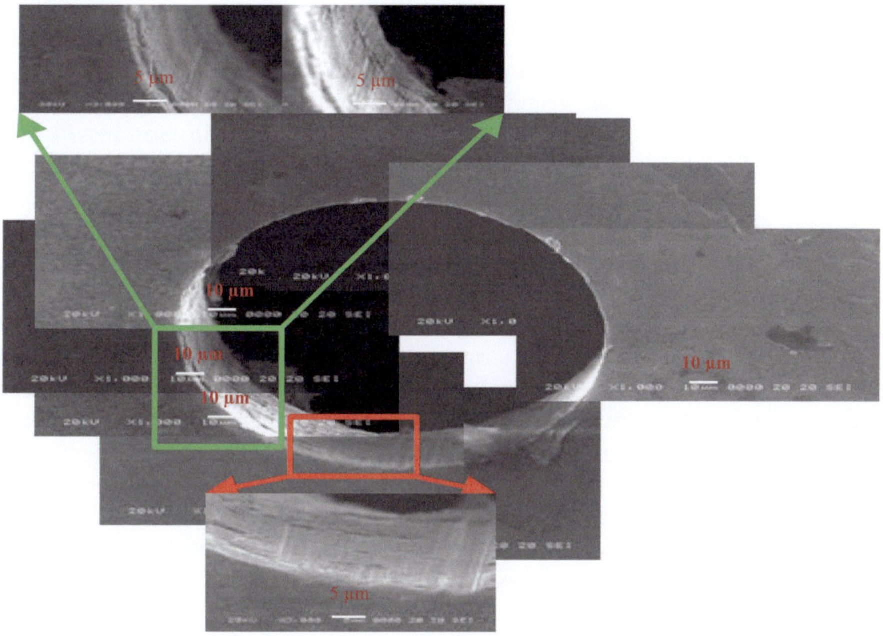

Fig. 5.17 Features of micropunching

The results mentioned in the previous sections reveal that the interaction between the micropunch and workpiece at the micro scale is significantly different from that in punching at the millimeter and macro scales. Consequently, a deeper exploration of the punch-foil interaction mechanism is necessary to advance the understanding of the relevant theory, ultimately leading to significant improvements in the quality of fabricated microholes for practical applications.

5.4 An Approach to Extending the Service Life: MWCNT Coating on Micropunches

Micropunching is a process in which a micropunch is pushed through a material and into a die to create a hole with the desired quality. However, as demonstrated in the experimental results given in the previous section, tool wear is a critical challenge that must be urgently addressed to maintain the efficiency and precision of the process. Enhancing the quality and durability of micropunches is a key strategy to mitigate wear loss, ultimately ensuring their successful application in various micromanufacturing scenarios.

Current findings suggest that even when micropunches are fabricated from tough and hard materials such as WC/Co, their morphology undergoes significant alterations as punching shots increase [28–34], particularly beyond 1525 shots. Due to friction between the micropunch and the substrate, the combined effects of severe Co depletion (as a bonding material) and the easy detachment of WC particles lead to substantial changes in micropunch morphology. This, in turn, results in a degraded profile of the punched microholes. Moreover, WC/Co micropunches are expensive [35], making their frequent replacement a costly challenge. Therefore, improving or optimizing the working conditions—particularly by reducing friction between the micropunch and the substrate—is essential for extending the service life of micropunches.

Carbon nanotubes (CNTs) have emerged as a highly attractive material in the field of tribology due to their exceptional structural and mechanical properties. Composed entirely of carbon (C) atoms arranged in a strong sp^2 C–C bonded network, CNTs exhibit a unique geometric structure that contributes to their outstanding mechanical strength, thermal stability, and electrical conductivity. Their high aspect ratio, elastic modulus, and superior tensile and flexural strengths make them particularly advantageous for various applications. Since their discovery [36, 37], CNTs have found widespread use in diverse fields, including ultrasensitive chemical and biosensors, field emission sources, photodetectors, transparent conductive membranes, nonvolatile memory, and electronics. Moreover, they play an increasingly vital role in reducing friction and wear rates in metal composites [38, 39], ceramics [40, 41], and nanotube-polymer composites [42, 43]. Theoretically, the coefficient of friction between the walls of multiwalled carbon nanotubes (MWCNTs) can be extremely low [44–50], making them a promising candidate for advanced tribological applications. Therefore, an MWCNT coating applied to the surface of the micropunch can effectively reduce friction between the micropunch and the substrate due to its graphitic nature, acting as a lubricant. This, in turn, mitigates wear and significantly extends the service life of the micropunch, ensuring improved performance and durability in micropunching applications.

Fig. 5.18 MWCNT coating on the micropunch

5.4.1 MWCNTs Coating on the Micropunch

Before the coating process, the micropunches were cleaned using acetone and pure ethyl alcohol to remove contaminants that could interfere with MWCNT growth and deposition. Iron (Fe) was used as the catalyst and was deposited onto the micropunch working area through electron cyclotron resonance (ECR) deposition. The processing parameters for ECR were as follows: vacuum pressure of 1.5×10^{-4} Pa, accelerated voltage of 2500 V, ion current density of 12.0 mA/cm^2, irradiation time of 60 s, with argon (Ar) as the gas and a flow rate of 0.6 SCCM. To prevent MWCNTs from forming on unintended areas, aluminum foil was applied to mask these regions [51, 52]. Following this, waste alcohol (complying with the GB-HW06 standard) was used as the carbon source for chemical vapor deposition (CVD) of MWCNTs. The CVD parameters included a vacuum pressure of 1×10^{-2} Pa, reaction direct current of 38–40 A, reaction time of 10 s, a heating rate of 3–4 A/s, and a cooling rate of 5 A/s.

Figure 5.18 shows a micropunch coated with MWCNTs along with corresponding high-magnification images. The images indicate that MWCNTs have successfully coated the micropunch surface to a height of approximately 15 μm. Moreover, the MWCNTs are vertically aligned with high density and exhibit a wiggly appearance due to the tightly packed structure formed during growth. Transmission electron microscopy (TEM) results further confirm that the MWCNTs have an approximate diameter of 3–5 nm, as shown in Fig. 5.19.

5.4.2 Wear Loss of the Micropunch Coated with MWCNTs

Figure 5.20 shows the relationship between the wear loss of the MWCNT-coated micropunch and the number of punching shots. The changes in the microtexture of the MWCNT-coated micropunch surface at various stages (excluding the final severe period) are also presented. The results show that the wear characteristics of the MWCNT-coated micropunch are similar to those of the uncoated micropunch. However, the quasi-stable period begins at approximately 400–450 punching shots,

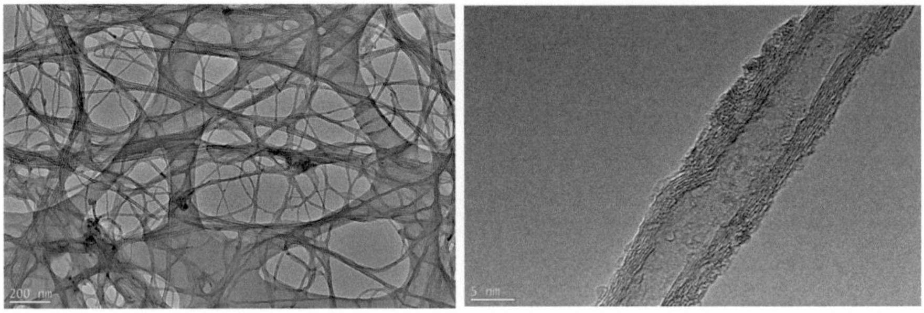

Fig. 5.19 TEM image of the MWCNTs

advancing by 50–100 shots compared to the uncoated micropunch. More notably, the quasi-stable period extends up to 1400–1425 punching shots, delaying its end by approximately 200–225 shots compared to the uncoated micropunch. The extended range of punching shots is also highlighted in the chart, demonstrating the effectiveness of the MWCNT coating in prolonging the lifespan of the micropunch. In the light of these results, it is possible to mention that the effect of the MWCNT coating on micropunch wear is evidently favorable. The quasi-stable period is significantly extended: for MWCNT-coated micropunches, this period ranges from approximately 400–450 shots to 1400–1425 shots, whereas for uncoated micropunches, it spans from 500 to 1200 shots. This extension effectively increases the service life of the micropunch by at least 250–325 shots. These findings confirm that MWCNT-coated micropunches exhibit greater durability compared to their uncoated counterparts, highlighting the beneficial impact of MWCNTs on the tribological performance of micropunches.

Comparing Fig. 5.14 with Fig. 5.20, the wear loss variation between MWCNT-coated and MWCNT-uncoated micropunches exhibits noticeable differences, despite following a similar overall trend. In the initial period, the wear characteristics of the MWCNT-coated micropunch show two peaks and one valley, indicating that its morphology remains relatively stable due to the lubricating effect of MWCNTs. In contrast, the uncoated micropunch experiences severe fluctuations in wear loss, with four peaks and three valleys, highlighting greater instability. Furthermore, in the severe period, although both micropunch types exhibit clear wear progression, the wavy variation in wear loss observed in earlier periods disappears, giving way to a steady increase—except for a fluctuation at 1550 punching shots. These findings affirm the robust effectiveness of MWCNT coating in mitigating wear loss, particularly during the quasi-stable period, significantly enhancing the durability and performance of the micropunch. Furthermore, the total wear loss of the MWCNT-coated micropunch is significantly lower than that of the MWCNT-uncoated micropunch, confirming that the MWCNT coating effectively reduces wear. This result clearly demonstrates that MWCNTs enhance the tribological performance of the micropunch, leading to a notable extension of its service life.

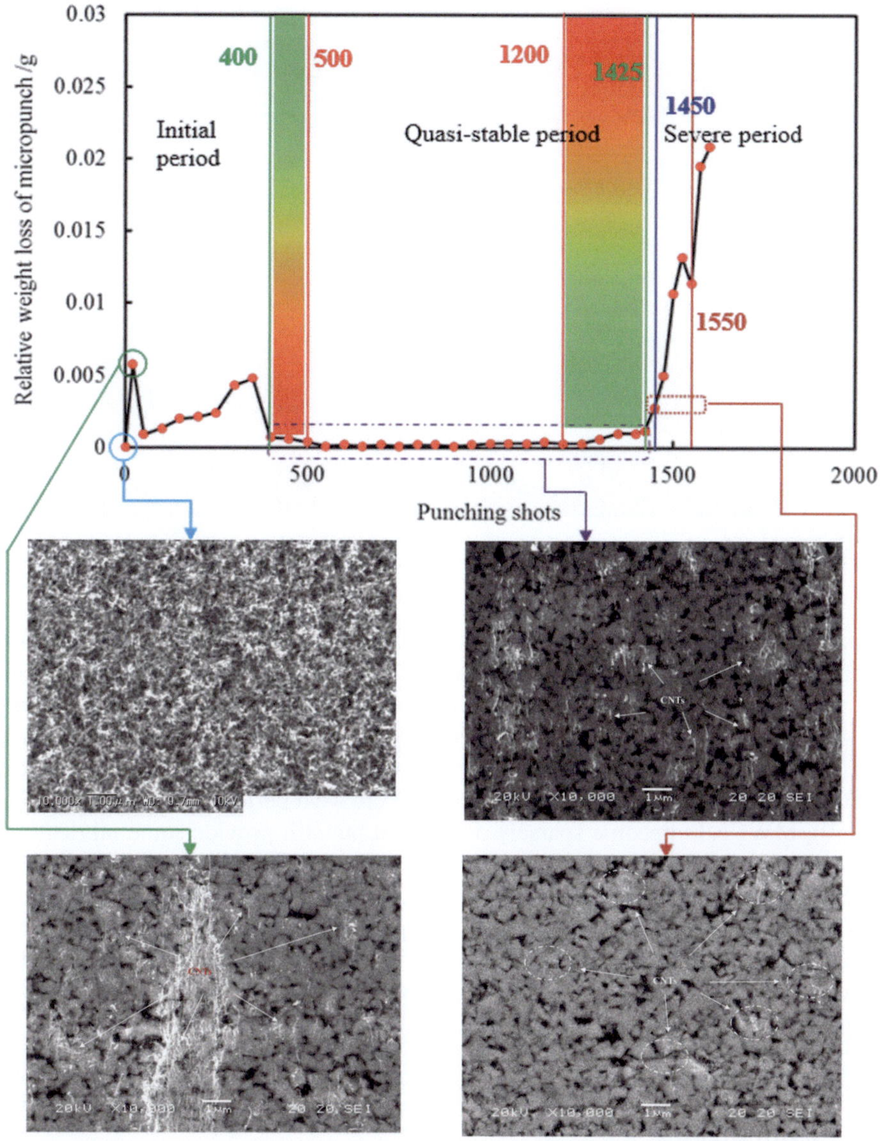

Fig. 5.20 Relationship between wear loss and punching shots for the MWCNT-coated micropunch

In addition to the wear loss results, SEM images depicting the morphological evolution of the MWCNT-coated micropunch are shown in Fig. 5.20, except for the final usage stage (i.e., at 1600 punching shots). Initially, from the top view of the coated micropunch, MWCNTs form a dense, tangled forest that is uniformly distributed across the working region. As punching progresses (e.g., at 50 shots in the run-in period), the rubbing effect between the micropunch and the substrate

Fig. 5.21 Microtexture of MWCNT-coated micropunch in the last severe wear period

causes bundles or clusters of MWCNTs to distribute nonuniformly on the micropunch surface. During the quasi-stable period, MWCNTs redistribute uniformly and act as an effective lubricant, reducing friction and promoting the formation of a lubrication layer or transfer film. This prevents direct contact between the micropunch and the substrate, thereby extending the service life of the micropunch in practical applications. As the micropunch reaches the severe wear period (e.g., punching shots >1450), MWCNTs become sparsely distributed but still provide some lubrication, ensuring that wear loss does not abruptly increase. However, at 1600 punching shots, the SEM image presented in Fig. 5.21 shows no remaining MWCNTs, indicating the loss of lubrication effects. At this stage, WC particles become the dominant feature on the surface, resembling the severely worn morphology observed in Fig. 5.11. Consequently, the micropunch undergoes rapid wear, leading to a sharp increase in wear loss.

5.5 Conclusions

In microtechnology at the micrometer scale, the miniaturization, integration, and enhanced functionality of components and systems remain the ultimate goals. Microcomponents have the potential to provide precise functionalities in various fields, including biomedical applications and sensor technologies. In particular, patterned microholes are promising candidates for lab-on-chip technology, where specific functionalized surfaces regulate analytical functionalities. While macrolevel techniques have been adapted for downsizing hybrid microsystem assemblies, there is an urgent need for highly integrated microsystems with greater precision. This demand is driven by the growing need for low-cost, high-quality, and miniaturized products across various industries, including consumer electronics, household appliances, and manufacturing. Consequently, new approaches are continuously being explored to meet these evolving requirements [53–61]. To date, microdevices are primarily made from glass or silicon. Micromachining, micromolding, and photoresist techniques are commonly employed for fabrication. However, in these traditional fabrication methods, the substrate is inevitably affected by the applied energy sources (such as heating, melting, vaporizing, and solidifying), which leads to unavoidable variations in the substrate properties. As a result, the final microdevices may not align well with practical requirements. Moreover, harmful chemicals are often used in the microfabrication process, posing environmental risks. Micropunching addresses the key issues associated with traditional technologies, minimizing potential environmental and human health risks. Furthermore, micropunching is a highly promising and environmentally friendly technique, as it does not require chemicals, offers high efficiency for large-scale production, and reduces costs. Moreover, the heat-affected zone (HAZ) is eliminated, and the debris generated during micropunching can be recycled. These advantages highlight that micropunching contributes to reducing waste, minimizing the depletion of natural resources, and ensuring that all material and energy inputs and outputs are as safe and sustainable as possible.

In this study, the near-net shape microholes were successfully formed using a WC/Co micropunch with a 150 μm diameter. During the initial phase of the micropunching process (punching shots <500), the wear of the micropunch increases significantly, with the primary cause of wear loss being the easy peeling-off of the Co component. As a result, the size of the micropunch remains unstable. When the punching shots range from 500 to 1200, the size of the micropunch stabilizes, as the wear condition of the micropunch enters a quasi-stable phase with minimal wear loss. During this period, the formed microholes exhibit high quality, making them ideal for near-net shape fabrication in micromanufacturing. However, as the punching shots exceed 1525, the significant wear of the micropunch leads to a noticeable decrease in its size, accompanied by wear loss from both Co and WC. Consequently, the micropunch becomes more unstable, and the quality of the microholes significantly declines. Moreover, the characteristics of microholes produced by micropunching differ considerably from those produced by conventional punching. In the case of a 200 μm thick titanium workpiece with a 150 μm diameter punched microhole, the areas of rollover and burr are much narrower than in conventional punching. Moreover, the fracture band is absent. Even in the severe wear period, the diameter of the front side of the punched microholes is nearly identical to that of the back side. However, the quality of the punched holes deteriorates substantially after approximately 1200 punching shots, due to the severe wear of the micropunch. To extend the service life of micropunches, MWCNT coating was successfully applied to the working region, resulting in a dense, tightly packed, and tangling structure with a diameter of approximately 3–5 nm and a height of about 15 μm. The method used for synthesizing the MWCNTs is eco-friendly, as waste alcohol was utilized as the carbon source. Although the wear characteristics of MWCNT-coated micropunches are similar to those of micropunches without MWCNT coating, the lubricating effect of the MWCNTs forms a protective layer or transfer film at the contact region. This film prevents direct contact between the micropunch and the substrate, thanks to the graphitic nature of the MWCNTs, and significantly extends the service life (quasi-stable period) by approximately 250–325 punching shots, which represents about 35.7–46.4% more shots compared to uncoated micropunches.

Based on these promising results, the development of techniques for producing strong and long-lasting micropunches is crucial and urgently needed. At the same time, the underlying mechanisms should be thoroughly investigated to enable practical applications with robust wear resistance. In addition, further exploration of environmentally friendly and green techniques, with minimal potential environmental impacts and risks to human health, is highly recommended.

References

1. Dornfeld D, Min S, Takeuchi Y. Recent advances in mechanical micromachining, *CIRP Annals - Manufacturing Technology* 2006; 55(2): 745–768.
2. Mark JJ. Microfabrication and nanomanufacturing. Boca Raton FL CRC/Taylor & Francis 2006.
3. Murali A, Lokhande G, Deo KA, Brokesh A, Gaharwar AK. Emerging 2D nanomaterials for biomedical applications. *Materials Today* 2021; 50: 276–302.
4. Dinis A, Mendes PM. A comprehensive review of powering methods used in state-of-the-art miniaturized implantable electronic devices. *Biosensors and Bioelectronics* 2021; 172: 112781.
5. Knobloch H, Kaminorz Y. MicroNano Integration. Springer-Verlag Berlin Heidelberg 2004.
6. Mohamed GH. The MEMS handbook. Boca Raton FL CRC/Taylor & Francis 2006.
7. Edwards SA. The nanotech pioneers. *Journal of Angewandte Chemie* 2006; 45(43): 7116–7117.
8. Sundaram S, Chen CS. Next-generation engineered microsystems for cell biology: a systems-level roadmap. *Trends in Cell Biology* 2022; 32(6): 490–500.
9. Sagathiya N, Sharma V, Ramkumar J. Electrochemical turning operation: A state of the art review. *Journal of Manufacturing Processes* 2024; 113: 131–151.
10. Mamalis AG. Advanced manufacturing engineering. *Journal of Materials Processing Technology* 2005; 161: 1–9.
11. Janssen PJM, Hoefnagels JPM, de Keijser ThH, Geers MGD. Processing induced size effects in plastic yielding upon miniaturization. *Journal of the Mechanics and Physics of Solids* 2008; 56: 2687–2706.
12. Uhlmann E, Mullany B, Biermann D, Rajurkar KP, Hausotte T, Brinksmeier E. Process chains for high-precision components with micro-scale features. *CIRP Annals - Manufacturing Technology* 2016; 65: 549–572.
13. Korkmaz E, Gozen BA, Bediz B, Ozdoganlar OB. Accurate measurement of micromachining forces through dynamic compensation of dynamometers. *Precision Engineering* 2017; 49: 365–376.
14. Cole KM, Kirk DW, Singh CV, Thorpe SJ. Optimizing electrochemical micromachining parameters for Zr-based bulk metallic glass. *Journal of Manufacturing Processes* 2017; 25: 227–234.
15. Biasetti DA, Di Liscia EJ, Torchia GA. Optical waveguides fabricated in Cr:LiSAF by femtosecond laser micromachining. *Optical Materials* 2017; 73: 25–32.
16. Singh A, Manikandan NA, Sankar MR, Pakshirajan K, Roy L. Experimental investigations and surface morphology of bio-micromachining on copper. *Materials Today: Proceedings* 2018; 5(2), Part 1: 4225–4234.

17. Saxena KK, Qian J, Reynaerts D. A review on process capabilities of electrochemical micromachining and its hybrid variants. *International Journal of Machine Tools and Manufacture* 2018; 127: 28–56.
18. Guo KW. Chapter 9 - Approaches to microholes for fabrication of microdevices. *Micro- and Nanotechnology Enabled Applications for Portable Miniaturized Analytical Systems. Micro and Nano Technologies* 2022; 197–216.
19. Gupta K, Laubscher RF. Sustainable machining of titanium alloys: A critical review. *Proceedings of the Institution of Mechanical Engineers Part B: Journal of Engineering Manufacture* 2017; 231(14): 2543–2560.
20. Geiger M, Kleiner M, Eckstein R, Tiesler N, Engel U. Microforming. *CIRP Annals* 2001; 50(2): 445–462.
21. Raulea LV, Goijaerts AM, Govaert LE, Baaijens FPT. Size effects in the processing of thin metal sheets. *Journal of Materials Processing Technology* 2001; 115(1): 44–48.
22. Engel U. Tribology in microforming. *Wear* 2006; 260(3): 265–273.
23. Qin Y. Micro-forming and miniature manufacturing systems - development needs and perspectives. *Journal of Materials Processing Technology* 2006; 177(1–3): 8–18.
24. Lowrie J, Ngaile G. New punch design for the elimination of punch ejection load through manipulation of the elastic strain field in the punch nose. *Journal of Manufacturing Processes* 2016; 22: 49–59.
25. Janča A, Siegl J, Haušild P. Small punch test evaluation methods for material characterization. *Journal of Nuclear Materials* 2016; 481: 201–213.
26. Ngaile G, Lowrie J. Punch design for floating based micro-tube hydroforming die assembly. *Journal of Materials Processing Technology* 2018; 253: 168–177.
27. Liu T, Wang Y, Hong M, Venezuela J, Shi W, Dargusch M. Advances in biodegradable piezoelectrics for medical implants. *Nano Today* 2023; 52: 101945.
28. Oliver G, Henning K, Pieter T. Microsystem engineering of lab-on-a-chip devices. *Angewandte Chemie* 2004; 43(34): 4382–4390.
29. Johnstone RW, Parameswaran M. An introduction to surface-micromachining. Kluwer Academic Publishers Netherlands 2004.
30. Guo KW. The wear characteristics of WC/Co micropunch. *Micro and Nanosystems* 2009; 1(3): 205–209.
31. Brousseau EB, Dimov SS, Pham DT. Some recent advances in multi-material micro- and nano-manufacturing. *International Journal of Advanced Manufacturing Technology* 2010; 47: 161–180.
32. Bhattacharyya B. Chapter 12-Recent advancements in EMM for micro and nanofabrication. *Electrochemical Micromachining for Nanofabrication, MEMS and Nanotechnology* 2015; 219–240.
33. Luttge R. Chapter 2-Basic technologies for microsystems. *Nano- and Microfabrication for Industrial and Biomedical Applications* (Second Edition) 2016; 11–54.
34. Guo KW, Tam HY. Ti microholes potential for biomedical application punched by 75WC/25Co micropunch. *Journal of Applied Biotechnology & Bioengineering* 2017; 4(3): 572–576.
35. Guo KW, Tam HY. Effects of extended punching on wear of the WC/Co micropunch and the punched microholes. *The International Journal of Advanced Manufacturing Technology* 2011; 59: 955–960.
36. Iijima S. Helical microtubules of graphitic carbon. *Nature* 1991; 354: 56–58.
37. Iijima S. Single-shell carbon nanotubes of 1-nm diameter. *Nature* 1993: 363: 603–605.
38. Bakshi SR, Keshri AK, Agarwal A. A comparison of mechanical and wear properties of plasma sprayed carbon nanotube reinforced aluminum composites at nano and macro scale. *Materials Science and Engineering: A* 2011; 528: 3375–3384.
39. Guiderdoni Ch, Pavlenko E, Turq V, Weibel A, Puech P, Estournès C, Peigney A, Bacsa W, Laurent Ch. The preparation of carbon nanotube (CNT)/copper composites and the effect of the number of CNT walls on their hardness, friction and wear properties. *Carbon* 2013; 58: 185–197.

40. Zhan GD, Kuntz JD, Wan JL, Mukherjee AK. Single-wall carbon nanotubes as attractive toughening agents in alumina-based nanocomposites. *Nature Materials* 2003; 2(1): 38–42.
41. Hvizdoš P, Puchý V, Duszová A, Dusza J, Balázsi C. Tribological and electrical properties of ceramic matrix composites with carbon nanotubes. *Ceramics International* 2012; 38: 5669–5676.
42. Green MJ, Behabtu N, Pasquali M, Adams WW. Nanotubes as polymers. *Polymer* 2009; 50: 4979–4997.
43. Pöllänen M, Pirinen S, Suvanto M, Pakkanen TT. Influence of carbon nanotube-polymeric compatibilizer masterbatches on morphological, thermal, mechanical, and tribological properties of polyethylene. *Composites Science and Technology* 2011; 71: 1353–1360.
44. Damnjanović M, Vuković T, Milosević I. Super-slippery carbon nanotubes: symmetry breaking breaks friction. *European Physical Journal B* 2002; 25: 131–134.
45. Dickrell PL, Sinnott SB, Hahn DW, Raravikar NR, Schadler LS, Ajayan PM, Sawyer WG. Frictional anisotropy of oriented carbon nanotube surfaces. *Tribology Letters* 2005; 18(1): 59–62. https://doi.org/10.1007/s11249-004-1752-0.
46. Dickrell PL, Pal SK, Bourne GR, Muratore C, Voevodin AA, Ajayan PM, Schadler LS, Sawyer WG. Tunable friction behavior of oriented carbon nanotube films. *Tribology Letters* 2006; 24(1): 85–90.
47. De Volder MF, Tawfick SH, Baughman RH, Hart AJ. Carbon nanotubes: Present and future commercial applications. *Science* 2013; 339: 535–539.
48. Guo KW, Tam HY. Effects of carbon nanotubes on wear of WC/Co micropunches. *The International Journal of Advanced Manufacturing Technology* 2014; 72: 269–275.
49. Zhai P, Isaacs JA, Eckelman MJ. Net energy benefits of carbon nanotube applications. *Applied Energy* 2016; 173: 624–634.
50. Guo KW. Carbon nanotubes coating for micropunch. *Journal of Coating Science and Technology* 2020; 9: 11–19.
51. Cortés-López AJ, Muñoz-Sandoval E, López-Urías F. Efficient carbon nanotube sponges production boosted by acetone in CVD-synthesis. *Carbon* 2018; 135: 145–56.
52. Rather SU. Preparation, characterization and hydrogen storage studies of carbon nanotubes and their composites: A review. *International Journal of Hydrogen Energy* 2020; 45(7): 4653–72.
53. Romig Jr. AD, Dressendorfer PV, Palmer DW. High performance microsystem packaging: A perspective. *Microelectronics Reliability* 1997; 37(10–11): 1771–1781.
54. Razzacki SZ, Thwar PK, Yang M, Ugaz VM, Burns MA. Integrated microsystems for controlled drug delivery. *Advanced Drug Delivery Reviews* 2004; 56(2): 185–198.
55. Wise KD. Integrated sensors, MEMS, and microsystems: Reflections on a fantastic voyage. *Sensors and Actuators A: Physical* 2007; 136(1): 39–50.
56. Fu WB, Turcheniuk K, Naumov O, Mysyk R, Wang FJ, Liu M, Kim DY, Ren XL, Magasinski A, Yu MH, Feng XL, Wang ZL, Yushin G. Materials and technologies for multifunctional, flexible or integrated supercapacitors and batteries. *Materials Today* 2021; 48: 176–197.
57. Gojun M, Šalić A, Zelić B. Integrated microsystems for lipase-catalyzed biodiesel production and glycerol removal by extraction or ultrafiltration. *Renewable Energy* 2021; 180: 213–221.
58. Manvi M, Swamy KBM. Microelectronic materials, microfabrication processes, micromechanical structural configuration based stiffness evaluation in MEMS: A review. *Microelectronic Engineering* 2022; 263: 111854.
59. Kulkarni MB, Goel S. Recent advancements in integrated microthermofluidic systems for biochemical and biomedical applications - A review. *Sensors and Actuators A: Physical* 2022; 341: 113590.
60. Juska VB, Maxwell G, Estrela P, Pemble ME, O'Riordan A. Silicon microfabrication technologies for biology integrated advance devices and interfaces. Biosensors and Bioelectronics 2023; 237: 115503.
61. Kulkarni MB, Yashas, Vyas R. A role of integrated microheaters in a microfluidics based point-of-care-testing and beyond for healthcare applications. *Applied Materials Today* 2024; 38: 102225.

Chapter 6
Micromanufacturing of Ultrathin Metal Foils via Rolling Processes

Ning Nie, Huijun Li, and Lihong Su

Abstract This chapter reviews the key factors influencing the production of ultrathin materials through rolling processes, including roll configurations, rolling force, and material properties. Beyond conventional rolling methods, innovative techniques such as asymmetrical rolling, Sendzimir mill rolling, double rolling, and accumulative pack rolling offer potential solutions to surpass theoretical thickness limits. Moreover, the chapter discusses the mechanical properties of ultrathin materials, including extreme extensibility and size effects, while presenting the underlying mechanisms driving these phenomena.

Keywords Ultrathin metal foils · Rolling processes · Asymmetrical rolling · Sendzimir mill · Double rolling · Accumulative pack rolling

6.1 Introduction

The continuous pursuit of advanced materials with exceptional performance, reliability, and precision is a key driver of innovation in modern manufacturing industries. As technology rapidly evolves, the demand for high-performance materials capable of meeting stringent functional and structural requirements is steadily increasing. One such class of materials is ultrathin metallic foils, which are characterized by their exceptionally small thicknesses, often measured in micrometers. These foils play a crucial role in a diverse range of cutting-edge applications across multiple industries, including flexible electronics [1, 2], where they contribute to the development of lightweight, bendable circuits; energy storage devices [3, 4], where they enhance battery efficiency and performance; aerospace engineering [5, 6], where their lightweight yet durable nature supports advanced structural and functional components; and nuclear physics research [7, 8], where their precision and stability are essential for experimental and shielding purposes. Beyond their small dimensions,

N. Nie · H. Li · L. Su (✉)
School of Mechanical, Materials, Mechatronic and Biomedical Engineering, University of Wollongong, Wollongong, NSW, Australia
e-mail: lihongsu@uow.edu.au

S. Gürgen (ed.), *Micromanufacturing Operations*,
https://doi.org/10.1007/978-3-032-04193-7_6

ultrathin metallic foils must also exhibit carefully controlled microstructural properties, including grain size, crystallographic texture, and mechanical strength, to ensure optimal functionality in their respective applications. Achieving such precise characteristics necessitates advanced manufacturing techniques capable of delivering both dimensional accuracy and surface integrity. Among the various micromanufacturing methods available, rolling processes stand out as the most efficient, cost-effective, and scalable solution for producing ultrathin foils with superior surface finishes, mechanical properties, and microstructural control. The rolling process allows manufacturers to produce large volumes of ultrathin metallic foils while maintaining tight thickness tolerances and achieving the required material performance characteristics, making it an indispensable technique in the production of high-performance metallic materials.

This chapter discusses the micromanufacturing of ultrathin materials through rolling processes. It begins with the fundamentals of ultrathin foil fabrication, examining the relationship between process parameters and foil characteristics. The discussion provides an overview of the principles governing rolling operations, with a focus on the unique challenges posed by micrometer-scale dimensions, including achievable minimum thickness, surface/interface roughening, strain softening, and size effects. By integrating theoretical insights with practical methodologies, this chapter equips readers with a comprehensive understanding of rolling-based micromanufacturing for ultrathin foils.

6.2 Fundamentals of Rolled Ultrathin Metal Foils

Metallic foils are defined as metal sheets with thicknesses typically up to 300 μm [9, 10]. These materials are characterized by their flexibility, high surface-to-volume ratio, and excellent mechanical properties. While thickness classifications vary by industry, ultrathin foils are generally considered those below 100 μm, with certain applications requiring thicknesses under 10 μm [11, 12]. Achieving such dimensions typically relies on rolling processes to ensure consistent thickness, uniform microstructure, and minimal surface defects.

6.2.1 Significance of Rolled Metal Foils

Metal foils are widely used in microsystems, microelectromechanical systems (MEMS), and battery technologies, all of which hold significant market value. By 2023, revenue from sensors and actuators was projected to surpass $100 billion [13]. In the medical field, MEMS technologies facilitate point-of-care diagnostics and early disease detection [14]. The lithium-ion battery (LIB) industry primarily serves two markets: consumer electronics and electric vehicles, with the combined market expected to exceed $69 billion by 2022 [15]. Many of these products, including battery current collectors and micromanufactured mechanical components, are manufactured using ultrathin metal foils. The demand for thinner foils continues to rise to enable smaller and lighter devices, particularly in the battery sector, where current collector foils have been reduced to approximately 10 μm, as seen in products like Samsung's 30Q LIB [16]. This reduction in thickness aims to enhance energy density in batteries.

6.2.2 Minimum Achievable Thickness in Rolling Processes

Rolling processes are among the most widely used methods for fabricating ultrathin foils due to their scalability and efficiency. However, achieving extremely thin dimensions through rolling presents significant challenges. The minimum achievable thickness in rolling is governed by factors such as the roll gap, material flow stress, frictional forces, and the tension applied [12, 17, 18]. Typically, the minimum achievable thickness ($h_{\min}$) through rolling can be estimated using the Stone formula [17], as shown in Eq. 6.1.

$$h_{\min} = 3.58\frac{D\mu\left(1.15\sigma - \sigma_p\right)}{E}, \tag{6.1}$$

where D is the roll diameter, μ is the Coulomb friction coefficient between the roll and the sheet, E is the Young's modulus of the roll, σ is the yield strength of the sheet, and σ_p is the mean tensile stress.

In practice, the material undergoes significant plastic deformation as it is compressed between the rolls. However, metals also experience elastic deformation before they begin to plastically deform. At extremely small thicknesses, elastic deformation becomes more significant relative to plastic deformation. As a result, the material resists further thinning, and elastic recovery (or springback) can cause challenges in achieving very thin foils [19, 20]. Furthermore, the rolls themselves experience flattening due to the forces applied during the process. This flattening alters the pressure distribution between the rolls and the metal, affecting the achievable thickness. If the rolls are not perfectly rigid, this flattening limits the pressure that can be applied at the contact points, thereby restricting the extent to which the material can be thinned [20, 21].

In theoretical considerations, there are regions where the metal moves either faster or slower than the rolls, known as slip zones. In the forward slip zone (F), the material moves in the same direction as the rolls, while in the backward slip zone (B), it moves in the opposite direction, as shown in Fig. 6.1. The rolling pressures for these two slip zones can be derived from the formulas given in Eqs. 6.2 and 6.3, while Eq. 6.4 provides the coefficient δ [22].

$$P_f = \frac{K}{\delta}\left[(\delta + 1)\left(\frac{h_x}{h}\right)^{\delta} - 1\right] \tag{6.2}$$

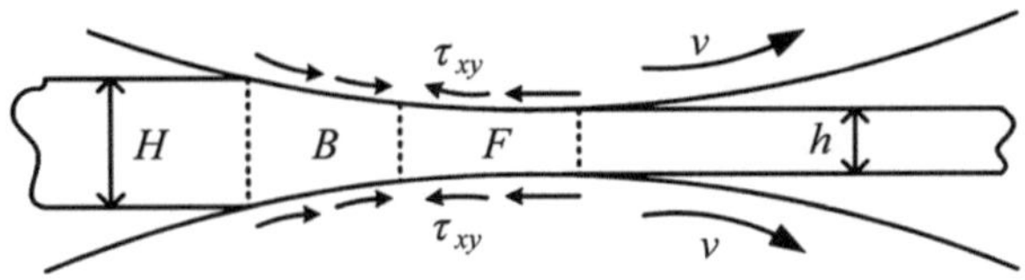

Fig. 6.1 Illustration of contact deformation zone in rolling [22]

$$P_b = \frac{K}{\delta}\left[(\delta - 1)\left(\frac{H}{h_x}\right)^{\delta} + 1\right] \tag{6.3}$$

$$\delta = \frac{2\mu l}{H - h} \tag{6.4}$$

where P_f and P_b are the rolling pressures applied in the forward slip zone and backward slip zone, respectively. K is the plane deformation, h_x is the foil thickness at any location along the contact arc, μ is the friction coefficient between the sheet and the rolls, l is the contact arc length, and h and H are the foil thicknesses at the exit and entrance, respectively.

These pressures result in a three-dimensional compressive stress state in the deformation zone. As the rolls flatten, the elongated arc length increases the peak pressure value. The center of the area will experience high hydrostatic pressure, as shown in Fig. 6.2, which prevents further thickness reduction [10, 22]. However, asymmetric rolling can create an additional cross-shear zone, which reduces the peak pressure at the center. This helps achieve a lower minimum achievable thickness.

Conventional rolling processes can achieve thicknesses as low as 20 μm [20], whereas specialized techniques, such as cluster rolling or asymmetrical rolling, can push this limit even further.

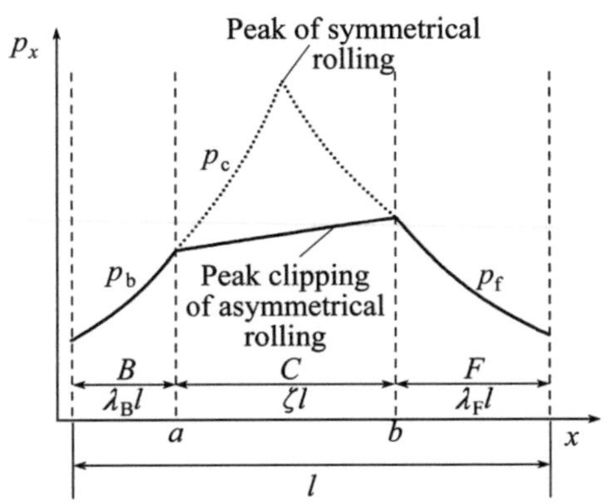

Fig. 6.2 Pressure distribution for asymmetric rolling: F is the forward slip zone, C is the cross-shear zone, and B is the backward slip zone [22]

6.2.3 Comparison Between Electrodeposited and Rolled Foils

Electrodeposition, as shown in Fig. 6.3, is an alternative method for producing ultrathin metallic foils [23–25], particularly for applications requiring extremely small thicknesses or specific compositions. While electrodeposited foils can achieve thicknesses well below 5 μm, they often exhibit different microstructural characteristics compared to rolled foils.

Electrodeposited foils generally feature equiaxed grain structures with relatively high levels of internal defects, such as voids and impurities. In contrast, rolled foils exhibit elongated grain structures that result from mechanical deformation and subsequent recrystallization, as shown in Fig. 6.4 [26, 27]. This microstructure gives rolled copper foils higher tensile strength but lower ductility compared to electroplated copper foils. The rolled foil exhibits a β-fiber texture, while the electroplated foil has a random crystal orientation. Furthermore, rolling processes are more suitable for large-scale production, being faster and more cost-effective than electrodeposition.

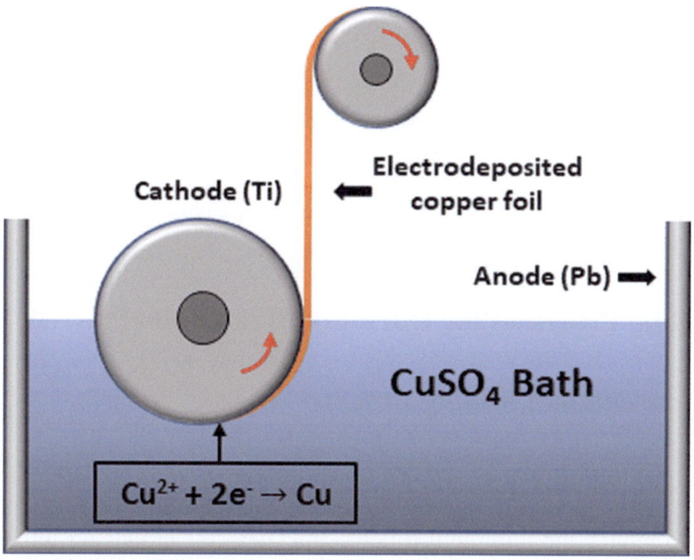

Fig. 6.3 Fabrication of copper foil by electrodeposition process

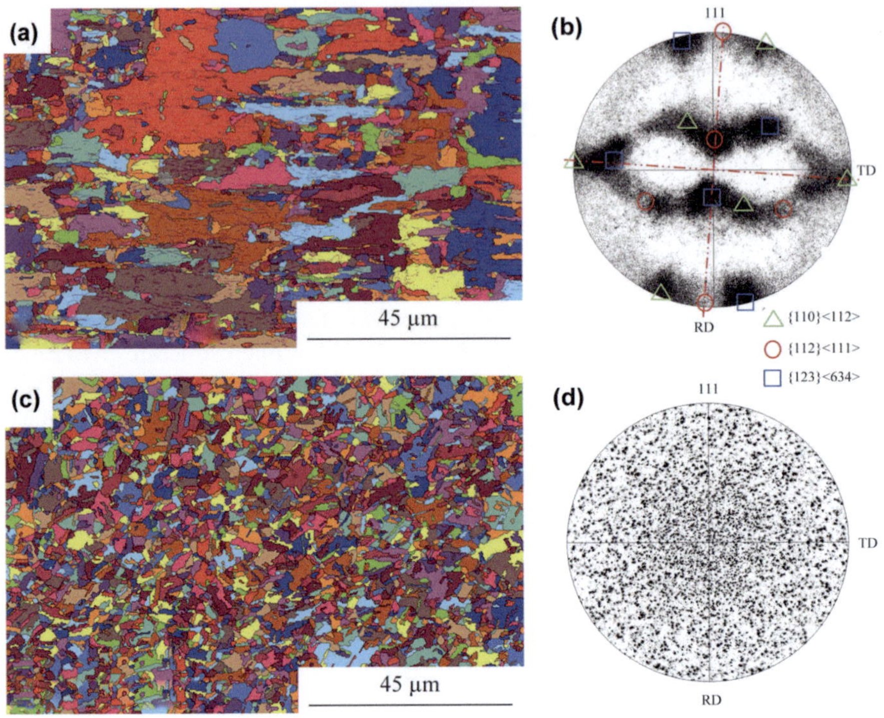

Fig. 6.4 Microstructure of copper foil: **a** rolled foil, **c** electrodeposited foil; and their <111> pole figure: **b** rolled foil and **d** electrodeposited foil [26]

6.3 Rolling Processes for Ultrathin Metal Foils

6.3.1 Conventional Rolling Process

The traditional rolling process, known for its efficiency and cost-effectiveness, has remained a focal point of extensive research over the past decade, particularly in efforts to reduce the minimum rollable thickness for producing ultrathin foils. According to the Stone formula [17], typical strategies for minimizing rollable thickness include reducing the roll diameter and elastic modulus, lowering the yield strength of the sheet, and applying tension.

6.3.1.1 Rolling Force

Rolling force plays a crucial role in determining the achievable thickness in the production of ultrathin foils. Classical models, such as the Stone formula [17], estimate the minimum thickness under idealized conditions but fail to account for practical constraints imposed by rolling force. Two key concepts refine these predictions: (i) the theoretical limiting thickness, which represents the idealized minimum unaffected by rolling force, and (ii) the conditional limiting thickness, which reflects the practical limitations of mill capacity. Increased rolling force reduces the minimum thickness by enabling greater plastic deformation within the deformation zone.

The theoretical framework builds upon the work of several predecessors. Fleck and Johnson [11] introduced the concept of distinct elastic and plastic zones in ultrathin foil rolling, providing a more detailed understanding of the deformation zones. Hacquin [28] and Weisz-Patrault et al. [29, 30] further advanced the analysis by applying Hertzian contact theory to describe the pressure distribution in the neutral zone, which was later verified through finite element simulations and sensor-based experiments. Experiments conducted on various materials, including pure aluminum, SS304 stainless steel, and copper, confirm the theoretical predictions [20], as shown in Fig. 6.5. The results demonstrate that higher rolling forces lead to thinner foils, approaching the theoretical limit. However, the extent of thickness reduction varies with material properties. Low yield strength materials, such as aluminum, can achieve a minimum thickness close to the theoretical value, while high yield strength materials, like stainless steel, require significantly higher forces but still fall short of the theoretical limits.

6.3.1.2 Rolling Speed

In foil rolling, the foil thickness generally decreases as the rolling speed increases. The reasons for this thinning effect can be summarized in three main points as given below.

(i) The rolling speed affects the deformation zone, frictional forces, and the metal flow rate during the rolling process. An increase in rolling speed can lead to a reduction in the coefficient of friction, thereby lowering the required rolling force. This reduction in rolling force is consistent with the experimental work of Jiang et al. [31]. Consequently, the rolling force reduction influences the final thickness of the rolled material, as shown in Fig. 6.6.

(ii) As the rolling speed increases, the two cylindrical rolls, which are loaded due to their interaction, tend to move closer together.

(iii) The temperature in the deformation zone increases significantly with the rise in rolling speed. As the material is bitten by the rolls, this temperature rise reduces its deformation resistance, resulting in work softening and making plastic deformation easier to achieve.

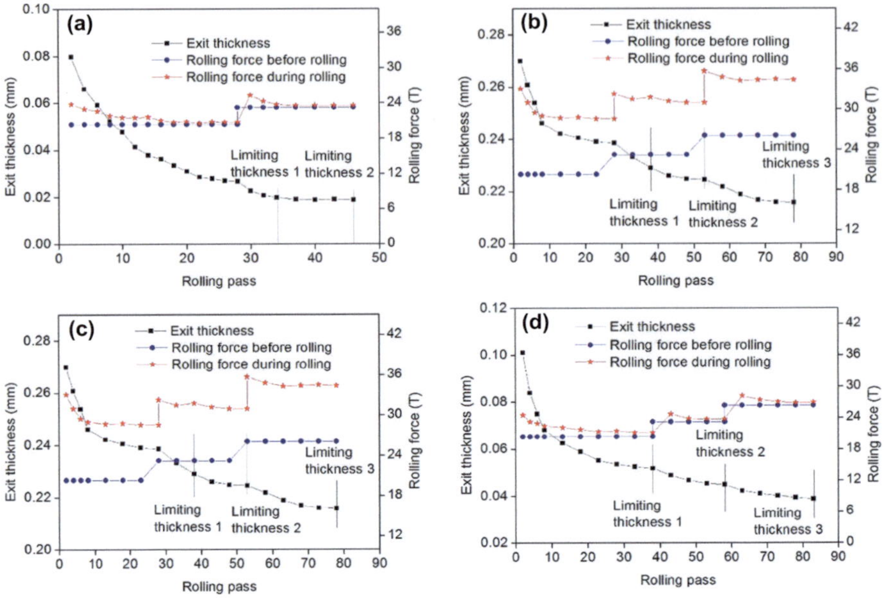

Fig. 6.5 The exit thickness obtained by given rolling force for different materials: **a** pure aluminum, **b** SS304 stainless steel, **c** nonannealed copper and **d** annealed copper [20]

Fig. 6.6 Effect of rolling speed on rolling force [31]

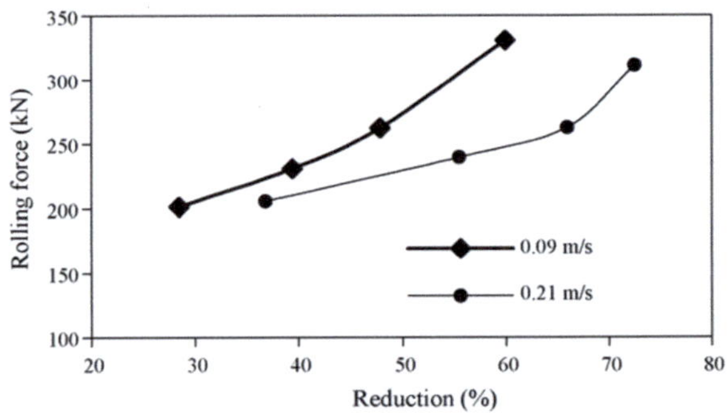

6.3.1.3 Rolling Speed Ratio and Asymmetric Rolling

Asymmetric rolling has emerged as an effective method for producing ultrathin foils. Unlike conventional rolling, which applies uniform forces and velocities to both rolls, asymmetric rolling involves differential velocities or roll diameters, creating a cross-shear zone in the deformation region [32, 33], as shown in Fig. 6.7. This cross-shear zone helps to reduce the minimum achievable thickness of rolled foils [10].

Early works have demonstrated the superiority of asymmetric rolling in overcoming the limitations of conventional rolling. For example, Yu et al. [34] produced copper foils as thin as 5 μm, a threshold that cannot be achieved through traditional

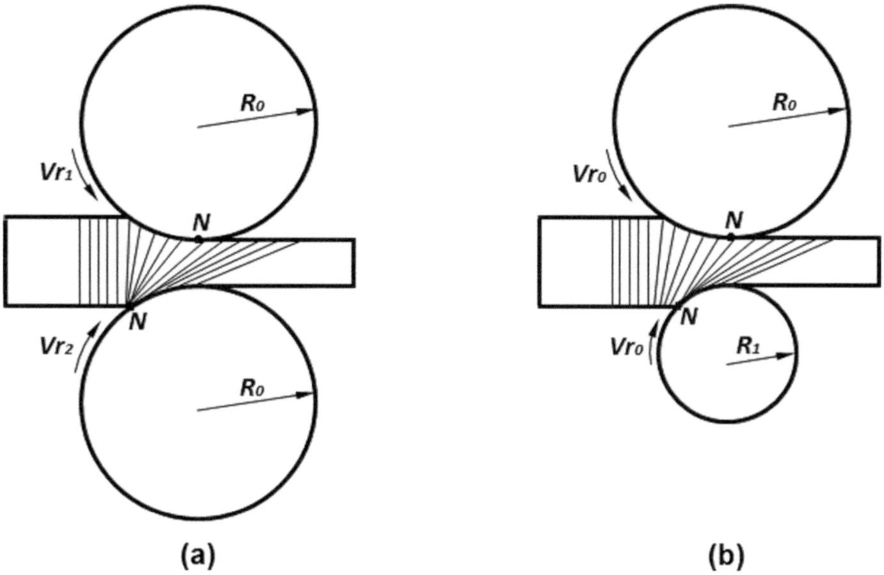

Fig. 6.7 Schematic illustration of asymmetric rolling: **a** rolling by different velocity of rolls and **b** rolling by different roll diameters; *N* is the neutral point

rolling. More recently, Tang et al. [10] empirically showed that asymmetric rolling could achieve a minimum thickness as low as 18 μm when the rolling speed ratio reached 1.3, which is approximately 30% of the thickness predicted by the Stone formula [17]. They also provided a new equation to predict the minimum achievable thickness for asymmetric rolling (with different roll velocities) as given in Eq. 6.5.

$$h_{\min} = (-1.3w^2 - 0.25w + 1.54)\mu CKR \tag{6.5}$$

where w is the proportion of the cross-shear zone in the deformation area, μ is the friction coefficient, C is the constant determined by Poisson's ratio and Young's modulus of the work roll and the rolled sheet, K is the plane deformation resistance, and R is the diameter of the roll.

The theoretical basis for reducing the minimum achievable thickness through asymmetric rolling lies in its unique deformation characteristics. In conventional rolling, the neutral plane, where frictional forces are balanced, limits the compressive stresses acting on the material, posing a barrier to further thinning. Asymmetric rolling disrupts this equilibrium by forming a cross-shear zone, as shown in Fig. 6.8, where opposing frictional forces on the upper and lower surfaces generate additional shear deformation. This reduces the hydrostatic stress required for plastic flow and allows for greater thickness reductions [10]. Furthermore, Liu et al. [22] incorporated elastic deformation into their model, demonstrating that asymmetric rolling achieves a lower minimum achievable thickness by redistributing stress along the deformation zone and minimizing the effects of elastic recovery.

Fig. 6.8 Deformation zone in asymmetrical rolling [19]

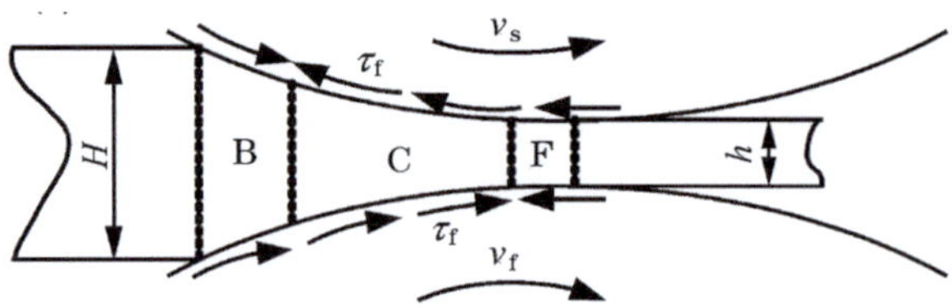

B—Backward-slip zone; C—Cross-shear zone; F—Forward-slip zone.

Fig. 6.9 Minimum thickness of foils obtained by different rolling speed ratios (i), and shear-cross region ratio (ω) of asymmetrical rolling [10]

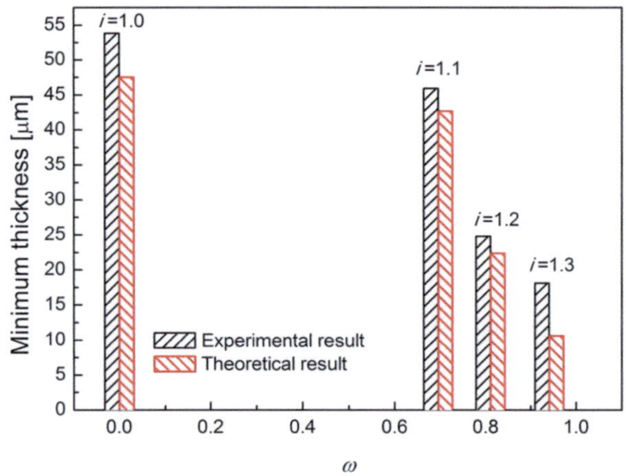

Key parameters influencing the minimum achievable thickness in asymmetric rolling have been well documented. It has been shown that higher speed ratios result in thinner foils [10]. This is directly linked to the expansion of the cross-shear zone as the speed ratio increases, as shown in Fig. 6.9. Similarly, Liu et al. [22] validated the minimum achievable thickness equation through experiments with asymmetry ratios ranging from 1.05 to 1.30. Their results revealed that higher asymmetry ratios, reduced roll diameters, and optimized friction coefficients enabled significant thickness reductions.

6.3.1.4 Friction and Lubrication

In cold rolling of thin and ultrathin metal foils, the friction and lubrication conditions play a significant role in determining the reduction in foil thickness. The lubricant not only reduces frictional forces but also influences the distribution of contact stress and the efficiency of the rolling process. As observed by Tateno et al. [12], lubricants with higher viscosities enhance thickness reduction by creating a more stable lubricant film between the rolls and the foil. Their experiments with mineral oils of varying

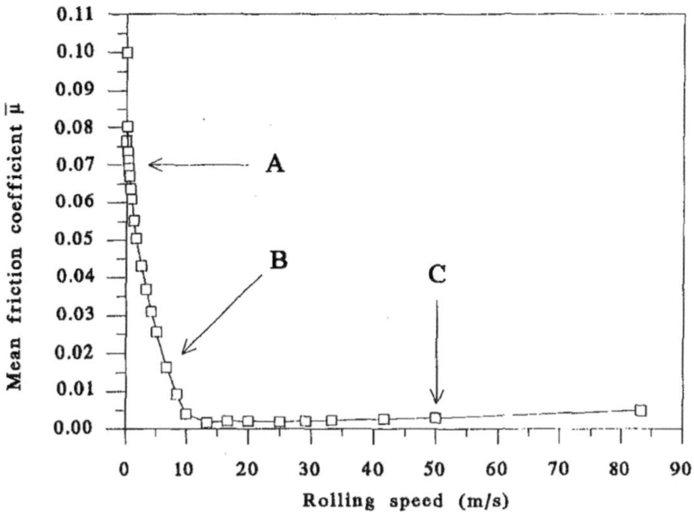

Fig. 6.10 Influence of rolling speed on mean friction coefficient; A is the boundary lubrication state, B is the hydrodynamic lubrication dominated state, and C is the completed hydrodynamic lubrication state [35]

viscosities demonstrated that thicker lubricants yielded higher reductions in thickness due to their ability to maintain mixed-film lubrication, minimizing direct contact between asperities on the roll and the foil surface.

Keife and Jonsater [35] suggested that rolling speed interacts with lubrication regimes, transitioning from boundary lubrication to mixed and hydrodynamic lubrication, as shown in Fig. 6.10. This transition directly impacts friction and rolling efficiency by reducing the mean friction coefficient, thereby enabling higher rolling reductions. However, they also cautioned that excessive rolling speed could lead to a fully hydrodynamic lubrication state, increasing friction and limiting further thickness reduction. Jiang et al. [31] also incorporated friction coefficients into predictive models, demonstrating their importance in estimating the minimum achievable thickness, particularly under conditions of elastic deformation and roll flattening. This highlights the importance of precise adjustments in roll surface conditions and rolling force, emphasizing their combined role in achieving significant thickness reductions while preventing surface defects such as edge cracks or excessive roll wear.

6.3.2 Cluster Rolling Process

The Sendzimir cluster rolling mill, as shown in Fig. 6.11, represents a significant advancement in cold rolling technology, particularly for producing ultrathin and high-precision metal foils. Invented in the 1930s [36], the Sendzimir mill features a 20-high

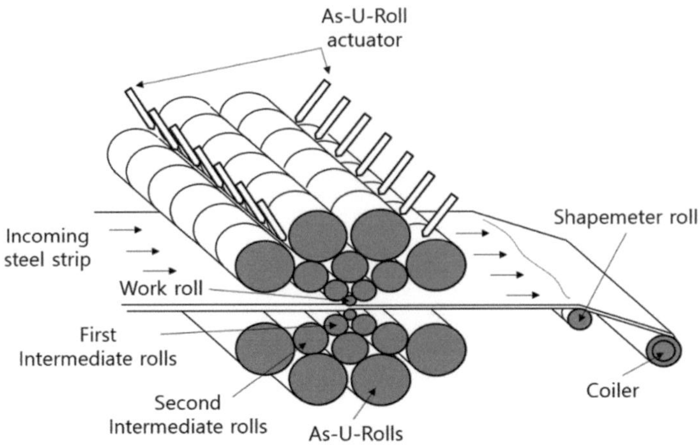

Fig. 6.11 Schematic illustration of the Sendzimir rolling mill [39]

roll configuration, utilizing small-diameter work rolls supported by intermediate and backup rolls. This design allows for better control of foil thickness and shape, making it ideal for processing challenging materials such as stainless steel and silicon sheets. Early research, such as Dutton's investigation [37], demonstrated the mill's capability to handle higher rolling loads while minimizing roll deflection—a key limitation in traditional four-high mills. As the demand for thinner foils continues to grow, roll diameters have steadily decreased. Currently, the Sendzimir rolling mill can achieve a minimum foil thickness of 1–5 μm [38].

One of the key technical features of the Sendzimir mill is its unique roll stack arrangement. The small-diameter work rolls are supported by intermediate rolls and multiple layers of backup rolls, which evenly distribute the rolling force across the foil. Moreover, advanced control mechanisms, such as hydraulic systems for adjusting roll gaps and As-U-Roll technology, enable precise control over foil flatness and thickness [39].

Numerical simulations conducted by Yu et al. [40] have demonstrated that adjustments in roll crown, taper, and tension significantly influence rolling process outcomes. For example, increasing the taper slope of the first intermediate roll can effectively reduce edge drop, ensuring uniform thickness across the foil width. Similarly, optimized tension and precise roll gap adjustments enable the production of ultrathin foils with minimal variation. These capabilities have solidified the role of the Sendzimir mill in industries requiring high-quality rolled products, such as electronics and automotive manufacturing. As research continues, advancements in modeling and control technologies are expected to further enhance the efficiency and capabilities of cluster rolling mills.

6.3.3 *Double Rolling Process*

Double rolling, as shown in Fig. 6.12, is a manufacturing process in which two metal sheets are stacked and rolled together to produce ultrathin foils, typically below 10 μm in thickness [41]. This method offers distinct advantages over single rolling, enabling the production of thinner foils while also doubling productivity, which would otherwise be limited by the time-intensive nature of single rolling [42, 43].

The minimum achievable thickness in double rolling depends on factors such as the mechanical properties of the material, roll dimensions, and the applied reduction. For example, double-rolled aluminum foils can reach thicknesses below 6 μm with minimal surface defects when optimal lubrication and rolling parameters are applied [45]. This process is crucial for industries requiring high-performance foils with ultrathin specifications and precise surface textures.

6.3.3.1 Surface Morphology of Rolled Foils

The double rolling process produces foils with two distinct surface morphologies. The external bright surface, which remains in contact with the polished rolls, is typically smooth with low roughness values. In contrast, the internal matte surface, formed at the interface between the two stacked sheets, exhibits a rough, wavy morphology [44]. This difference is attributed to localized deformation mechanisms, such as shear band formation, which create alternating peaks and valleys on the matte surface.

The roughness of the matte surface is a critical parameter in applications requiring strong adhesion, such as bonding with resin matrices in circuit boards. Double rolling allows for the control of surface roughness through adjustments in rolling reduction, lubrication, and material properties. For instance, studies on double-rolled copper foils indicate that the roughness of the matte surface can range from 1.0 to 1.3 μm, making it ideal for direct adhesion of resin in printed circuit board fabrication without the need for additional treatments [46].

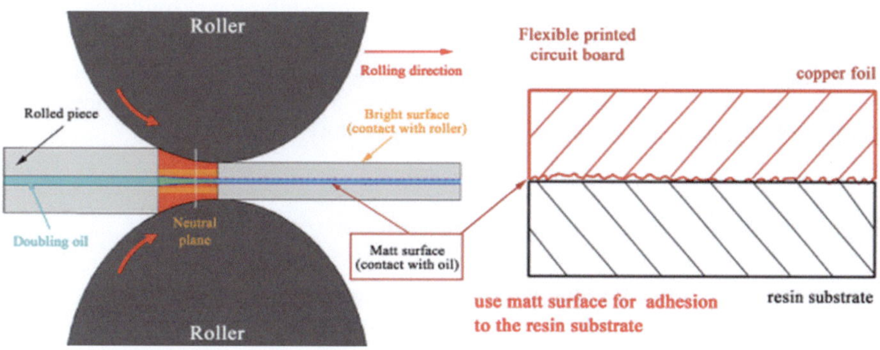

Fig. 6.12 Schematic illustration of double rolling process [44]

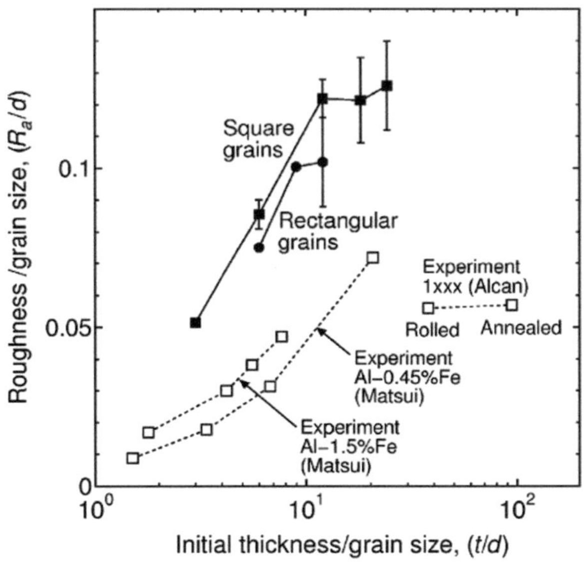

Fig. 6.13 Influence of the ratio of thickness to grain size on surface roughness in double rolling process [42]

6.3.3.2 Influence of Microstructure on Morphology

The microstructural characteristics of the metal, including grain shape, size, and distribution relative to the foil thickness, significantly influence the resulting surface morphology. Grain elongation in the rolling direction promotes more uniform deformation, resulting in smoother surfaces, whereas equiaxed grains contribute to increased surface roughness due to localized strain heterogeneities [47].

The ratio of grain size to foil thickness plays a crucial role in the double rolling process, as shown in Fig. 6.13. Foils with smaller grains relative to thickness undergo more uniform deformation, resulting in smoother matte surfaces [42, 48]. In contrast, larger grains promote pronounced shear banding, leading to increased surface roughness [46]. For example, studies on double-rolled aluminum foils indicate that materials with a fine-grained microstructure exhibit lower roughness amplitudes compared to those with coarser grains [49]. Furthermore, grain boundary characteristics and crystallographic orientations significantly influence deformation mechanisms. A higher grain boundary density in fine-grained foils restricts strain localization, leading to smoother surface textures [44].

6.3.3.3 Influence of Friction and Lubrication

The presence of lubricant at the interface between stacked aluminum sheets affects both surface roughness and sheet separation after rolling. Lubricants with lower viscosity, such as Somentor 32, form thinner films that promote smoother surfaces and better inter-sheet correlation. In contrast, thicker lubricants like Primol 352 tend

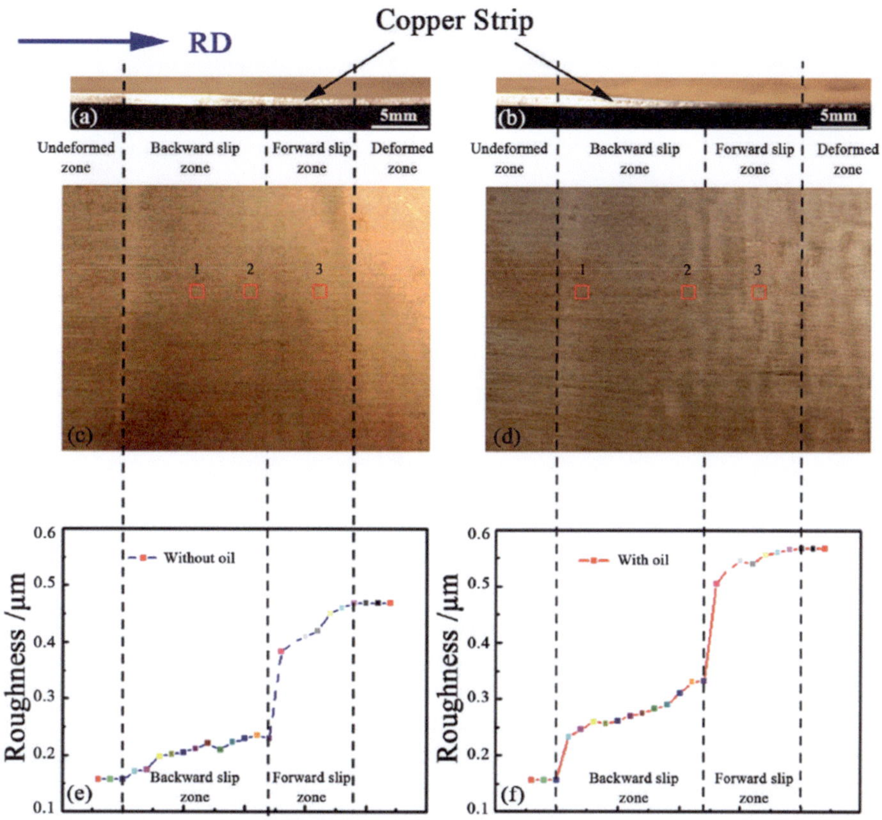

Fig. 6.14 Copper foil morphologies **a**, **b** on the surface of thickness cross-section, and **c**, **d** on the RD-TD surface with **e**, **f** roughness changes at corresponding regions [44]

to accumulate in surface valleys, slightly increasing roughness [50, 51]. Interestingly, the absence of lubricant often results in the smoothest surface but can complicate sheet separation. Experiments with copper foils, as shown in Fig. 6.14, have demonstrated that applied doubling oils significantly increase roughness while imparting specific surface characteristics [46]. In addition, friction at the roll surface impacts the smoothness of the bright surface.

6.3.4 Accumulative Pack Rolling Process

6.3.4.1 Principle of the Accumulative Pack Rolling

Accumulative pack rolling (APR) is an advanced technique developed to produce ultrathin metal foils through severe plastic deformation [52]. Unlike traditional

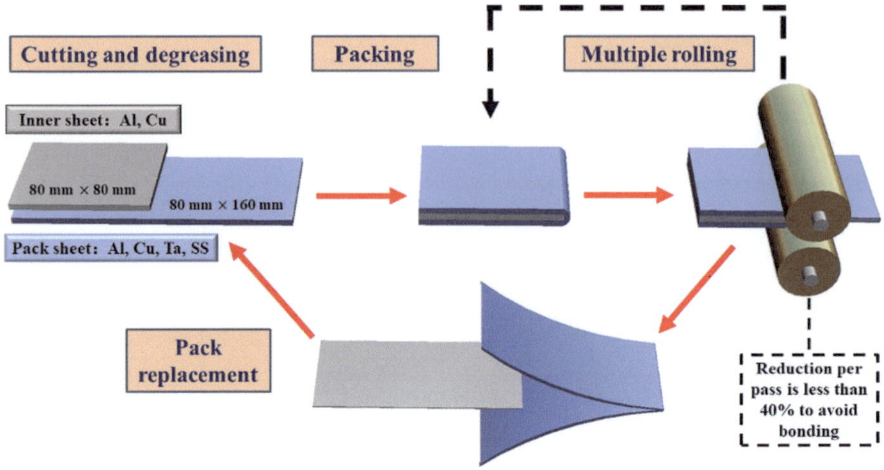

Fig. 6.15 Operational steps in double rolling process [52]

rolling methods, which face challenges in achieving thicknesses below a few micrometers due to limitations such as the elastic modulus of the rollers and friction, APR utilizes a sandwich structure. In this process, an inner sheet is enclosed between pack sheets made of materials like aluminum (Al), tantalum (Ta), or stainless steel (SS). This layered assembly is repeatedly rolled to achieve significant thickness reduction.

The process involves several critical steps, as shown in Fig. 6.15: (i) preparation of the sheets, (ii) cleaning of surfaces to avoid contamination, (iii) assembly of the pack structure, and (iv) multiple rolling passes. Each pass achieves a small thickness reduction, usually less than 20%, to prevent bonding or excessive deformation. After each significant accumulated reduction (above 60%), the pack sheets are replaced to maintain continuous processing. By repeating this cycle, APR can produce Al foils as thin as 2 μm, with surface and structural characteristics influenced by the chosen pack materials.

The selection of pack materials is crucial in determining the characteristics of the final foil. For example, Al/Al/Al packs are primarily used for basic applications, while Ta/Al/Ta and SS/Al/SS configurations improve surface quality and thickness uniformity. This innovative process not only allows Al foils to be reduced to thicknesses beyond the limitations of conventional rolling but also minimizes surface defects and plastic instability.

6.3.4.2 Surface Morphology of APR Processed Foils

During the APR process, the surface morphology of the foil undergoes dynamic evolution, influenced by both the pack materials and the degree of reduction, as shown in Fig. 6.16. Initially, the surface exhibits rolling marks parallel to the rolling direction. As the thickness reduction increases, these patterns gradually evolve into more

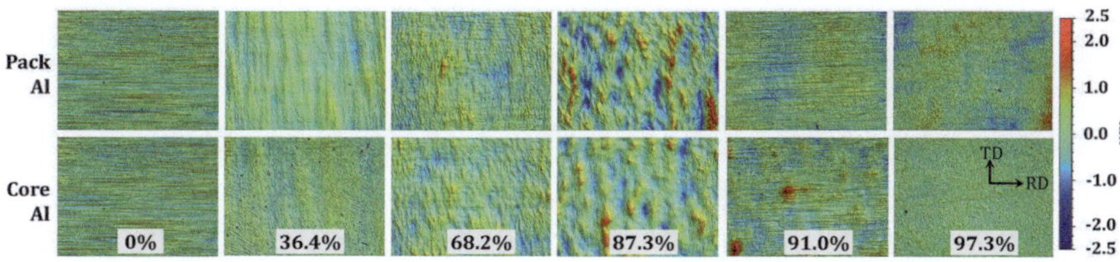

Fig. 6.16 Typical surface morphology change with the thickness reduction increase during the Al/Al/Al APR process [52]

complex roughening mechanisms, including ridging and particle-like roughening. At moderate reductions, ridging roughening develops, characterized by alternating ridges and valleys perpendicular to the rolling direction. These features become more pronounced as rolling progresses, leading to significant surface unevenness. However, beyond certain reductions (e.g., above 87.3%), the introduction of a fresh pack sheet can suppress ridging, resulting in a smoother surface. At high reductions, nearing 97.3%, particle-like roughening takes over, creating a uniformly distributed fine texture.

In the experimental results, the Al/Al/Al configuration exhibits pronounced ridging roughening at intermediate stages, while Ta/Al/Ta and SS/Al/SS configurations effectively suppress these features. Ta pack sheets exhibit minimal ridging due to their hardness and stability, while SS sheets help reduce surface roughening and maintain overall uniformity. This surface evolution enhances the functional properties of ultrathin foils, particularly for applications that require high-quality finishes.

6.3.4.3 Microstructure of APR Processed Foils

The microstructure of Al foils produced by APR plays a critical role in determining their mechanical and functional properties. Figure 6.17 shows the changes in microstructure from the initial Al sheet to the final foils fabricated using different pack materials. Initially, the Al sheet exhibits elongated, pancake-like grains aligned with the rolling direction, with high-angle grain boundaries (represented by black lines) surrounding them. After undergoing APR, the foil thickness is reduced to just a few grains, typically ranging from two to five. The grains are elongated along the rolling direction, though they become less pronounced compared to the initial sheet. High-angle grain boundaries dominate the structure, indicating significant plastic deformation. Low-angle grain boundaries (shown as grey lines) indicate localized deformation zones, which run parallel to the thickness direction. Equiaxed grains begin to emerge within the foil microstructure, particularly at the interfaces between elongated grains. These grains form as a result of subgrain rotation and the activation of multiple slip systems during deformation. The grain refinement and distribution achieved through APR significantly enhance the mechanical strength and ductility

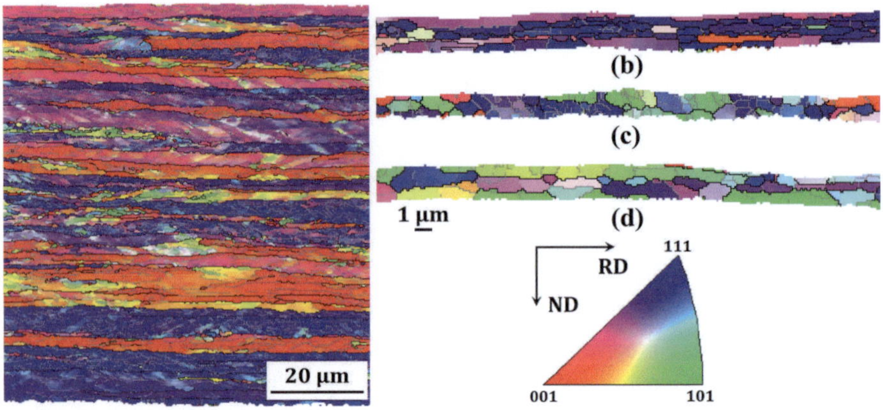

Fig. 6.17 Cross-section inverse pole figure map: **a** initial Al sheet and final Al foils fabricated by three configurations in three pack materials; **b** Al/Al/Al, **c** Ta/Al/Ta, **d** SS/Al/SS [52]

of the foil, making it well-suited for advanced applications that require ultrathin dimensions and uniform properties.

6.4 Mechanical Properties of Ultrathin Metal Foils

6.4.1 Extreme Extensibility of Rolled Metal Foils

Extreme extensibility, a remarkable mechanical property of rolled metal foils, has drawn significant attention due to its potential in advanced manufacturing. This phenomenon refers to the ability of materials, such as copper foil, to undergo exceptional deformation, achieving elongation well beyond conventional limits [53]. For instance, copper foils rolled under compound forming conditions have achieved extensibility ratios as high as 43,684%, as noted by Yu et al. [54]. This extraordinary capability is primarily attributed to a combination of mechanical and structural factors, including innovative rolling methods and microstructural evolution under strain.

Techniques like asymmetrical rolling, which generate shear deformation in addition to compression and tension, have been shown to activate multiple slip systems. This, in turn, enhances material ductility and reduces the risk of fracture. Furthermore, cold rolling at room temperature without intermediate annealing, as demonstrated by Lu et al. [53, 55], preserves strain energy and prevents grain growth. Moreover, the use of highly purified, dense nanocrystalline copper minimizes artifacts such as porosity, facilitating smooth deformation over extended rolling cycles. Contrary to the traditional understanding of strain hardening in metals, extreme extensibility gives rise to a phenomenon known as strain softening. In this case, material hardness initially increases with deformation but eventually peaks and then declines. Experimental observations, including those by Yu et al. [54], indicate that strain softening begins when the foil thickness drops below 100 μm. At this stage, the dislocation density becomes saturated, and deformation transitions from lattice dislocation movement to grain boundary sliding and rotation. This reduction in hardness allows further deformation without causing fracture, which is crucial for achieving ultrathin foils.

The microstructural evolution during rolling plays a critical role in enabling extreme extensibility. In the initial stages of rolling, grains are elongated and compressed, forming lamellar structures aligned with the rolling direction. As the foil becomes thinner, however, grain refinement and reorganization take place. This process leads to an increase in grain boundary misorientation, transitioning from small-angle to large-angle boundaries [53]. These changes promote enhanced grain boundary activities such as sliding and rotation, which become the dominant mechanisms of plastic deformation at the nanoscale. Dislocation entanglement within grains decreases, giving way to grain-boundary-mediated mechanisms.

6.4.2 Size Effect of Rolled Foils

The size effect refers to the phenomenon where the mechanical behavior and properties of materials differ significantly from those observed at the macroscale when the material dimensions are reduced to the microscale. This effect becomes particularly significant when the foil thickness approaches the scale of the microstructure, such as the grain size, leading to changes in the material deformation mechanisms. For example, when the grain size is comparable to the foil thickness, surface grains play a dominant role in the deformation process due to their reduced constraints, resulting in behaviors such as lower flow stress and reduced strain hardening [56]. Moreover, the surface layer model proposed by Geiger et al. [57] explains that surface grains experience less resistance to deformation compared to interior grains, because they lack dislocation pile-ups at grain boundaries, thus influencing both the flow behavior and the fracture characteristics of the material.

Size effects play a crucial role in the deformation behavior of rolled foils, especially in uniaxial tension and deep drawing processes. In tensile testing, as the thickness-to-grain-size ratio (t_0/d_0) decreases, both yield stress and ultimate tensile stress decrease significantly, as shown in Fig. 6.18. This reduction reflects the diminishing contribution of internal grains to the material strength [58]. The same trend is observed in deep drawing, where thinner foils display higher anisotropy and reduced formability due to the heterogeneous deformation of individual grains [59]. Moreover, free surface roughening becomes more pronounced during microscale deformation, as noted by Meng and Fu [58], leading to nonuniform thickness and localized strain concentrations, which ultimately affect the overall deformation behavior of the material.

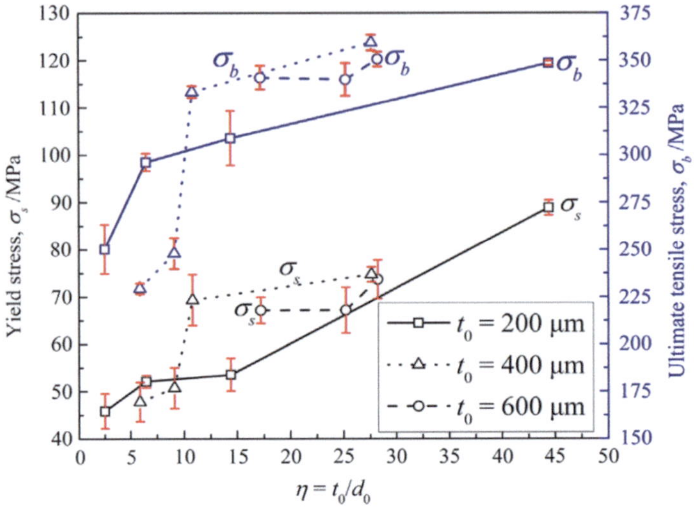

Fig. 6.18 Influence of thickness-to-grain-size ratio (t_0/d_0) on tensile stress [58]

Thinner foils tend to exhibit higher yield strength but reduced ductility, a trend attributed to the dominance of surface grains and the reduced dislocation activity in smaller structures [60, 61]. In fatigue testing, as shown in Fig. 6.19, the fatigue resistance of thin foils improves due to the suppression of microstructural defects such as voids and cracks, which are less likely to propagate in smaller geometries. In high-strain-rate micromechanical tests, size effects also manifest as varying responses in flow stress and fracture mechanisms. For instance, thinner foils exhibit higher strain rate sensitivity due to enhanced grain boundary activities, such as sliding and rotation [62]. These behaviors highlight the critical role of geometric and microstructural dimensions in determining deformation characteristics.

Fig. 6.19 Relations of fatigue life and applied strain amplitude ($\Delta\varepsilon/2$) of grain size constant copper foils [60]

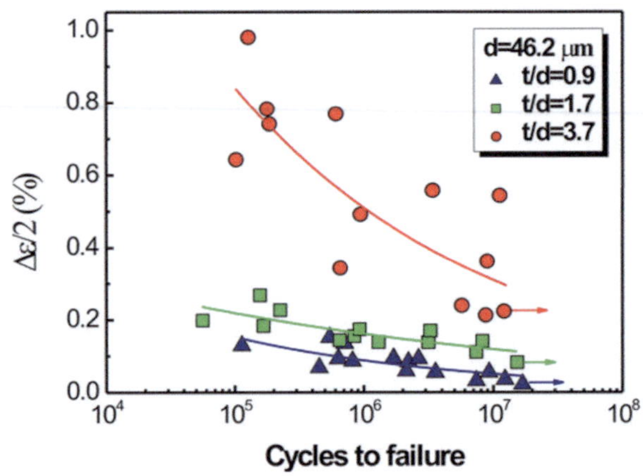

6.5 Summary

This chapter provides a comprehensive review of various rolling processes used to produce ultrathin metallic foils. These methods deal with key challenges such as dimensional precision, surface quality, and material properties while ensuring scalability and cost-effectiveness.

Traditional rolling is a widely used technique in foil production. According to the Stone formula, optimizing roll material, roll dimensions, and the properties of the rolled material can achieve smaller thicknesses. However, producing ultrathin foils below 10 μm remains challenging. Asymmetrical rolling, which utilizes work rolls with different diameters or varying roll speed ratios, presents a viable alternative. This approach introduces an additional cross-shear zone, enhancing deformation and enabling the production of thinner foils more efficiently. The Sendzimir mill is a sophisticated 20-roll system featuring two extremely small-diameter work rolls. These rolls are supported by backup rolls, minimizing work roll deflection and enabling the production of ultrathin foils. This design also allows for the rolling of materials with higher hardness. Double rolling and accumulative pack rolling utilize multilayer structures to efficiently produce ultrathin materials with enhanced productivity. Compared to single-layer rolling, these methods typically result in matt surface roughening at the contact interface between stacked sheets.

Rolling ultrathin sheets introduces unique performance characteristics, including extreme extensibility and strain softening. These effects are particularly evident in nanocrystalline materials, where an increase in high-angle grain boundaries and a reduction in the number of grains through the thickness direction influence deformation behavior. Moreover, the size effect is a key phenomenon in ultrathin foils, leading to a decrease in material strength as the ratio of foil thickness to grain size decreases. As the material becomes thinner, its ductility declines, while strain rate sensitivity increases. These properties enable rolled foils to meet the stringent requirements of modern applications, including electronics, energy storage, aerospace, and advanced manufacturing.

References

1. Jillek W, Yung W. Embedded components in printed circuit boards: a processing technology review. The International Journal of Advanced Manufacturing Technology. 2005;25:350–60.
2. Wang X, Liu Z, Zhang T. Flexible sensing electronics for wearable/attachable health monitoring. Small. 2017;13(25):1602790.
3. Chu H-C, Tuan H-Y. High-performance lithium-ion batteries with 1.5 μm thin copper nanowire foil as a current collector. Journal of Power Sources. 2017;346:40–8.
4. Yun Q, He YB, Lv W, Zhao Y, Li B, Kang F, et al. Chemical dealloying derived 3D porous current collector for Li metal anodes. Advanced materials. 2016;28(32):6932–9.
5. Solin JR. Shielding effectiveness of satellite faraday cages with EMI taped seams and closeouts. IEEE Electromagnetic Compatibility Magazine. 2018;7(2):40–6.

6. Asmatulu R, Bollavaram P, Patlolla V, Alarifi IM, Khan W. Investigating the effects of metallic submicron and nanofilms on fiber-reinforced composites for lightning strike protection and EMI shielding. Advanced Composites and Hybrid Materials. 2020;3:66–83.
7. Schwoerer H, Pfotenhauer S, Jäckel O, Amthor K-U, Liesfeld B, Ziegler W, et al. Laser-plasma acceleration of quasi-monoenergetic protons from microstructured targets. Nature. 2006;439(7075):445–8.
8. Robinson A, Zepf M, Kar S, Evans R, Bellei C. Radiation pressure acceleration of thin foils with circularly polarized laser pulses. New journal of Physics. 2008;10(1):013021.
9. Yu H, Lu C, Tieu AK, Godbole A, Su L, Sun Y, et al. Fabrication of ultra-thin nanostructured bimetallic foils by Accumulative Roll Bonding and Asymmetric Rolling. Scientific reports. 2013;3(1):2373.
10. Tang D, Liu X, Song M, Yu H. Experimental and theoretical study on minimum achievable foil thickness during asymmetric rolling. PloS one. 2014;9(9):e106637.
11. Fleck N, Johnson K. Towards a new theory of cold rolling thin foil. International journal of mechanical sciences. 1987;29(7):507–24.
12. Tateno J, Hiruta T, Katsura S, Honda A, Miyata T, Kamimaru A. Experimental analysis of thickness reduction limits in ultra thin stainless steel foil rolling. ISIJ international. 2011;51(5):788–92.
13. Hossain N, Al Mahmud MZ, Hossain A, Rahman MK, Islam MS, Tasnim R, et al. Advances of Materials Science in MEMS Applications: A Review. Results in Engineering. 2024:102115.
14. Nasreldin M, de Mulatier S, Delattre R, Ramuz M, Djenizian T. Flexible and stretchable micro-batteries for wearable technologies. Advanced Materials Technologies. 2020;5(12):2000412.
15. Zeng X, Li M, Abd El-Hady D, Alshitari W, Al-Bogami AS, Lu J, et al. Commercialization of lithium battery technologies for electric vehicles. Advanced Energy Materials. 2019;9(27):1900161.
16. Zhu P, Gastol D, Marshall J, Sommerville R, Goodship V, Kendrick E. A review of current collectors for lithium-ion batteries. Journal of Power Sources. 2021;485:229321.
17. Stone M. Rolling of thin strip. Iron Steel Eng. 1953;30:61.
18. Tong K, Sachs G. Roll-separating force and minimum thickness of cold-rolled strips. Journal of the Mechanics and Physics of Solids. 1957;6(1):35–46.
19. Tang D-l, Liu X-h, Li X-y, Peng L-g. Permissible minimum thickness in asymmetrical cold rolling. Journal of Iron and Steel Research International. 2013;20(11):21–6.
20. Xiao H, Ren Z, Liu X. New mechanism describing the limiting producible thickness in ultra-thin strip rolling. International journal of mechanical sciences. 2017;133:788–93.
21. Wusatowski Z. Fundamentals of rolling: Elsevier; 2013.
22. Liu X, Liu X-h, Song M, Sun X-k, Liu L-z. Theoretical analysis of minimum metal foil thickness achievable by asymmetric rolling with fixed identical roll diameters. Transactions of Nonferrous Metals Society of China. 2016;26(2):501–7.
23. Yin K, Xia Y, Chan C, Zhang W, Wang Q, Zhao X, et al. The kinetics and mechanism of room-temperature microstructural evolution in electroplated copper foils. Scripta Materialia. 2008;58(1):65–8.
24. Chang T, Jin Y, Wen L, Zhang C, Leygraf C, Wallinder IO, et al. Synergistic effects of gelatin and convection on copper foil electrodeposition. Electrochimica Acta. 2016;211:245–54.
25. Wang D-S, Song J-M, Lin S-C, editors. Mechanical properties of electrodeposited copper foils for PCB applications. 2010 5th International Microsystems Packaging Assembly and Circuits Technology Conference; 2010: IEEE.
26. Yin X-Q, Peng L-J, Kayani S, Cheng L, Wang J-W, Xiao W, et al. Mechanical properties and microstructure of rolled and electrodeposited thin copper foil. Rare Metals. 2016;35:909–14.
27. Merchant H, Liu W, Giannuzzi L, Morris J. Grain structure of thin electrodeposited and rolled copper foils. Materials characterization. 2004;53(5):335–60.
28. Hacquin A. 3D thermomechanical modelling of rolling processes: coupling strip and rolls: PhD Dissertation, Ecole des Mines de Paris, 1996 (in French); 1996.
29. Weisz-Patrault D, Ehrlacher A, Legrand N. A new sensor for the evaluation of contact stress by inverse analysis during steel strip rolling. Journal of Materials Processing Technology. 2011;211(9):1500–9.

30. Weisz-Patrault D, Maurin L, Legrand N, Salem AB, Bengrir AA. Experimental evaluation of contact stress during cold rolling process with optical fiber Bragg gratings sensors measurements and fast inverse method. Journal of Materials Processing Technology. 2015;223:105–23.
31. Jiang Z, Wei D, Tieu A. Analysis of cold rolling of ultra thin strip. Journal of Materials Processing Technology. 2009;209(9):4584–9.
32. Vincze G, Simões FJ, Butuc MC. Asymmetrical rolling of aluminum alloys and steels: a review. Metals. 2020;10(9):1126.
33. Simões FJP. Asymmetrical rolling of an aluminum alloy 1050: Universidade de Aveiro (Portugal); 2008.
34. Jiuming Y, Guangfeng J, Quan Z. The deformation characteristics in cross shear cold rolling of ultra-thin strip and the theory of elastic stopper. J of Northeastern University. 1982;3(3):17–27.
35. Keife H, Jonsater T. Influence of rolling speed upon friction in cold rolling of foils. 1997.
36. Yu H-L, Liu X-H, Lee G-T, Park H-D. Numerical analysis of strip edge drop for Sendzimir mill. Journal of materials processing technology. 2008;208(1–3):42–52.
37. Dutton K. An investigation into the design and performance of an automatic shape control system for a Sendzimir cold rolling mill: Sheffield Hallam University (United Kingdom); 1983.
38. White J. Cold rolling of sheet and strip: aluminium and aluminium alloys. Metals Technology. 1975;2(1):346–53.
39. Seo M, Ban J, Koo BY, Kim SW. Static model identification for sendzimir rolling mill using noise corrupted operation data. IEEE Access. 2020;8:150685–95.
40. Yu H-L, Liu X-H, Lee G-T, Park H-D. Numerical analysis of strip edge drop for Sendzimir mill. Journal of Materials Processing Technology. 2008;208(1):42–52.
41. Larsson F. Surface topography measurements of thin aluminium foil transferred to FE-simulations. 2016.
42. Utsunomiya H, Sutcliffe M, Shercliff H, Bate P, Miller DJIJoMS. Evolution of matt surface topography in aluminium pack rolling. Part I: model development. 2004;46(9):1349–64.
43. Utsunomiya H, Sutcliffe M, Shercliff H, Bate P, Miller D. Three-dimensional topography of the matt surface in aluminium pack rolling. International journal of mechanical sciences. 2004;46(7):1083–96.
44. Wang X, Liu X, Shi L, Li J, Xie J. Characteristic and formation mechanism of matt surface of double-rolled copper foil. Journal of Materials Processing Technology. 2015;216:463–71.
45. Utsunomiya H, Sutcliffe M, Shercliff H, Bate P, Miller D. Influence of friction on roughening of the matt surface in aluminium pack rolling. International Journal of Machine Tools and Manufacture. 2005;45(7–8):803–9.
46. Wang X-y, Liu X-f, Zou W-j, Xie J-x. Copper foils with gradient structure in thickness direction and different roughnesses on two surfaces fabricated by double rolling. International Journal of Minerals, Metallurgy, and Materials. 2013;20(12):1170–5.
47. Utsunomiya H, Sutcliffe M, Shercliff H, Bate P, Miller D. Evolution of matt surface topography in aluminium pack rolling. Part II: effect of material properties. International journal of mechanical sciences. 2004;46(9):1365–75.
48. Matsui K. Mechanism of pinhole formation in aluminium foil rolling. Kobelco Technology Review. 1991;12(10):60–4.
49. Utsunomiya H, Sutcliffe MPF, Shercliff HR, Bate P, Miller DB. Evolution of matt surface topography in aluminium pack rolling. Part II: effect of material properties. International Journal of Mechanical Sciences. 2004;46(9):1365–75.
50. Utsunomiya H, Sutcliffe M, Shercliff H, Miller D. Experimental investigation of roughening of the matt surface in pack rolling. Journal of materials processing technology. 2006;177(1–3):501–4.
51. Kimura H, editor Discussion on pinhole formation in pack rolling of aluminium foils. Proceedings of 70th Spring Meeting Japan Institute of Light Metals, JILM, Tokyo; 1986.
52. Nie N, Wang R, Deng G, Wang H, Wang P, Tieu AK, et al. Fabrication of Ultrathin Al Foils by Accumulative Pack Rolling: Influence of Pack Materials. Metals. 2024;14(11):1262.

53. Lu L, Sui M, Lu K. Superplastic extensibility of nanocrystalline copper at room temperature. Science. 2000;287(5457):1463–6.
54. Yu Q, Liu X, Tang D. Extreme extensibility of copper foil under compound forming conditions. Scientific Reports. 2013;3(1):3556.
55. Lu L, Sui M, Lu K. Cold rolling of bulk nanocrystalline copper. Acta materialia. 2001;49(19):4127–34.
56. Simons G, Weippert C, Dual J, Villain J. Size effects in tensile testing of thin cold rolled and annealed Cu foils. Materials Science and Engineering: A. 2006;416(1–2):290–9.
57. Geiger M, Kleiner M, Eckstein R, Tiesler N, Engel U. Microforming. CIRP annals. 2001;50(2):445–62.
58. Meng B, Fu M. Size effect on deformation behavior and ductile fracture in microforming of pure copper sheets considering free surface roughening. Materials & Design. 2015;83:400–12.
59. Meng B, Wang W, Zhang Y, Wan M. Size effect on plastic anisotropy in microscale deformation of metal foil. Journal of Materials Processing Technology. 2019;271:46–61.
60. Dai C, Zhang B, Xu J, Zhang G. On size effects on fatigue properties of metal foils at micrometer scales. Materials Science and Engineering: A. 2013;575:217–22.
61. Chen JQ, Gao HT, Hu XL, Yang LQ, Ke DW, Liu XH, et al. The significant size effect on nucleation and propagation of crack during tensile deformation of copper foil: Free surface roughening and crystallography study. Materials Science and Engineering: A. 2020;790:139678.
62. Zhu C, Xu J, Yu H, Shan D, Guo B. Size effect on the high strain rate micro/meso-tensile behaviors of pure titanium foil. Journal of Materials Research and Technology. 2021;11:2146–59.

Chapter 7
Physical Vapor Deposition of Thin Film Coatings for Cutting Tools

Esad Kaya

Abstract Thin film coatings are widely applied to various products to enhance hardness, wear resistance, conductivity, and aesthetics. Industries such as aerospace, electronics, automotive, and machinery utilize thin film coatings to improve functional performance. These coatings are produced using microdeposition techniques, encompassing a wide range of processes. This chapter provides an overview of thin film deposition technology, with a particular focus on physical vapor deposition (PVD) techniques. It presents the fundamental principles and mechanisms of thin film formation using PVD. The chapter also discusses cutting tool coatings, focusing on key coating materials and their roles in machining applications.

Keywords Microdeposition techniques · Physical vapor deposition · Thin film · Coating · Cutting tools

7.1 Introduction

Thin film coating techniques have been widely used to modify material surfaces, particularly at microscale and nanoscale. These coatings play a crucial role in high-end applications such as aerospace, electronics, and the automotive industry. Several deposition techniques, including atomic layer deposition (ALD), chemical vapor deposition (CVD), and physical vapor deposition (PVD), have been developed to achieve precise control over surface properties [1]. Unlike conventional surface treatments, these advanced methods impart specialized functions—such as enhanced hardness, wear resistance, and conductivity—without significantly altering the bulk properties of the material. Thin film coatings are typically defined as materials with thicknesses ranging from a few nanometers to several micrometers. Alternatively, they can be described as layers thinner than one micrometer that extend freely in two dimensions while being constrained in the third [1, 2].

E. Kaya (✉)
Department of Mechanical Engineering, Eskişehir Osmangazi University, Eskişehir, Türkiye
e-mail: esatkaya@ogu.edu.tr

S. Gürgen (ed.), *Micromanufacturing Operations*,
https://doi.org/10.1007/978-3-032-04193-7_7

These structures are synthesized using nanotechnology and are classified as nanomaterials, which exhibit distinct properties compared to bulk materials due to their high surface area and quantum effects [3]. Thin films produced through deposition can consist of atoms, molecules, or ions as their fundamental building blocks [1, 2]. When the deposited material exceeds a thickness of one micrometer, it is classified as a thick film [1]. In the microstructural analysis of these materials, grain structures at both microscale and macroscale are observed, in contrast to the finer grains typically found in thin films [1, 2, 4]. The primary difference between thin and thick film categories lies in their deposition techniques. Thin film manufacturing processes are generally more complex, requiring high-vacuum environments, whereas thick film production relies on simpler, cost-effective methods [1, 3]. This distinction is evident in the manufacturing of metal resistors. Thin film resistors are fabricated using atomic-level techniques such as sputtering, whereas thick film resistors are produced using methods like screen printing and serigraphy [3]. In terms of performance, thin film techniques offer higher precision and better electrical properties, while thick film techniques provide enhanced durability and surface protection [2–4].

Various approaches are used for thin film deposition, many of which require high-voltage operation and a high-vacuum environment. These deposition processes create thin films on different substrates, with the surface receiving the deposition commonly referred to as the substrate. Thin films are applied through methods such as chemical reactions, sputtering, and thermal deposition techniques. The choice of deposition technique significantly affects the microstructure—such as impurity levels and void formation—as well as key performance criteria, including mechanical, optical, and chemical properties [1, 2, 4, 5]. Thin film deposition is governed by fundamental principles involving key components such as the target, precursor, substrate, and coating. The target refers to the primary material used to form a thin layer on a surface through deposition processes. The surface onto which the thin film is deposited is known as the substrate. For example, in the sputtering deposition of aluminum nitride thin films, pure solid aluminum serves as the target material, while nitrogen acts as the reactive gas. A precursor is typically a gaseous substance that interacts with other materials within the reaction chamber to facilitate thin film formation. Under vacuum conditions, the thin film material is deposited onto the substrate, ensuring precise control over film composition and structure [1, 6].

In CVD, PVD, and ALD, the substances used to achieve the desired chemical stoichiometry on surfaces are known as precursors. The substrate refers to the surface onto which thin or thick film layers are applied, with the primary goal of deposition techniques being to modify or enhance its properties. During the deposition process, a thin coating forms on the surface and continues to grow under controlled conditions [1]. Deposited thin film materials are commonly referred to as coatings. These coatings, whether thin or thick films, serve to protect the substrate surface while improving its properties. Typically, coatings enhance mechanical characteristics such as elastic modulus, hardness, and toughness, as well as tribological properties like friction and wear resistance. Furthermore, they improve the resistance to corrosion by altering chemical reactivity at the substrate surface [1, 2, 7, 8].

Traditional painting techniques are among the most commonly used methods for modifying material surfaces. Similar to conventional painting, thin film deposition and indirect coating techniques impart properties to a surface that the substrate material alone cannot achieve. These properties may include optical, electrical, or biocompatible characteristics, depending on the film composition and structure. Thin film coatings offer excellent resistance to high temperatures, humidity, and exposure to acidic or alkaline environments. Moreover, thin film deposition techniques are not only employed for functional improvements but also for decorative applications, enhancing both aesthetics and durability [1, 2, 6].

7.2 Thin Film Deposition Techniques

Coating materials can be applied to surfaces using several processes, which are typically categorized based on the source of deposition for the thin film material. The classification shown in Fig. 7.1 is widely recognized in the field. There are three fundamental coating methods: atomic-level deposition techniques (gaseous), particle deposition methods (solution-based), and molten state techniques [1].

The gas phase is predominantly utilized in thin film deposition processes, where the source material must first transition into a gaseous phase to enable coating formation. In this process, the source material is transformed into a gas, which then deposits

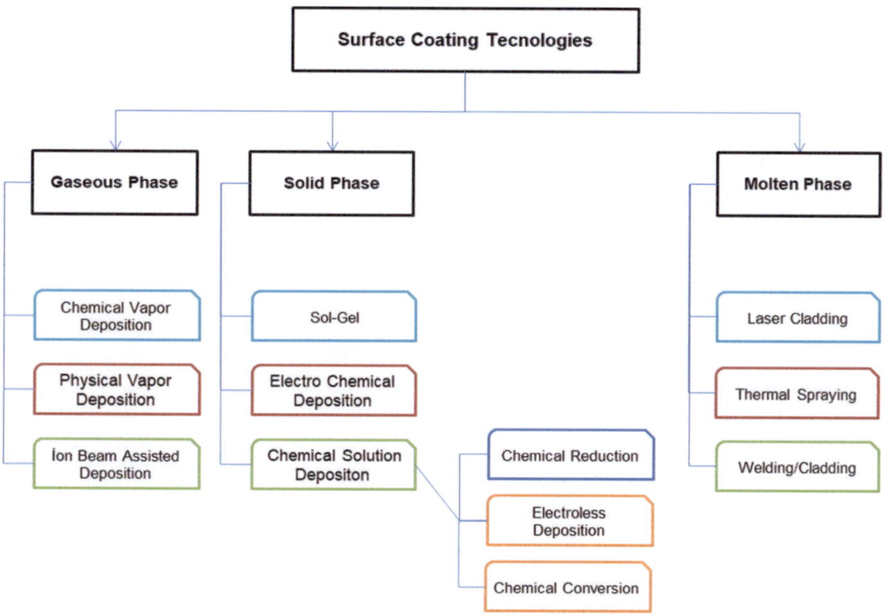

Fig. 7.1 Deposition techniques used in coating processes

on the surface to form the thin film. The primary techniques used in gas-phase deposition include PVD, CVD, and ion beam-assisted deposition technologies. In PVD, the source material is initially vaporized, and the evaporated material undergoes atomization before deposition on the substrate. This technique does not require a chemical reaction during the coating process. Variations of PVD include sputtering, thermal spraying, arc vapor deposition, ion beam deposition, and pulsed laser deposition, each offering unique modifications to the core PVD method [9]. CVD is another gas-phase thin film deposition technique, which involves chemical reactions. It includes various subtypes such as thermal CVD, low-pressure CVD, laser CVD, photon-assisted CVD, plasma-assisted CVD, and ALD. Advanced techniques like pulsed CVD/ALD and atomic layer epitaxy, due to their enhanced capabilities, are sometimes not classified as traditional CVD methods. CVD offers several advantages, including the ability to produce extremely thin films through chemical reactions, achieve conformal deposition with excellent uniformity, create hybrid nanostructures, and fabricate semiconductors [1, 6]. Thin films are formed using solution-based deposition processes that require liquids and solutions. In these processes, the precursor must be in solution form to generate the thin film. Examples of such techniques include chemical bath deposition, sol–gel processes, and electrochemical deposition. The underlying mechanism involves chemical reactions initiated by the chemical solutions, with precipitates forming from the reaction and being deposited on the substrate material [1, 10]. The source material must be converted into a liquid phase using molten deposition fabrication techniques. Methods such as laser melting, thermal spraying, and other source-dependent processes are commonly employed. In these techniques, the molten material is atomized and deposited onto the substrate under high pressure [1].

The growth behavior of thin films is primarily influenced by the characteristics of the substrate material. Moreover, it is directly affected by the layer thickness and the deposition technique used. Typically, the formation of a thin film involves three fundamental processes. The first step is the creation of the thin film, where the interaction among the precursor, reactants, and target material plays a crucial role. The second step involves transferring the material to the substrate, usually via a vacuum transfer process. In the third and final phase, the thin film structure develops on the substrate surface. Several parameters also influence the characteristics of the thin film deposited on the substrate. These include activation energy, bonding energy between the substrate and the thin film, and the adhesion strength between the two materials [1, 6, 9].

The nucleation mechanism of thin film production is classified into three primary categories, as shown in Fig. 7.2. Figure 7.2a shows the Frank-van der Merwe mechanism. This growth mechanism involves an imbalance in surface interactions, where the attraction between the particles adhering to the surface is stronger than the attraction between the surface and an individual particle. As a result, the thin film develops through the aggregation and coalescence of atoms on the surface, leading to the formation of islands [1]. Figure 7.2b shows the Volmer-Weber thin film growth method, which exhibits both island development and layered growth. Initially, a layer forms, and then islands begin to emerge on top of that layer [1, 11]. In Fig. 7.2c, the

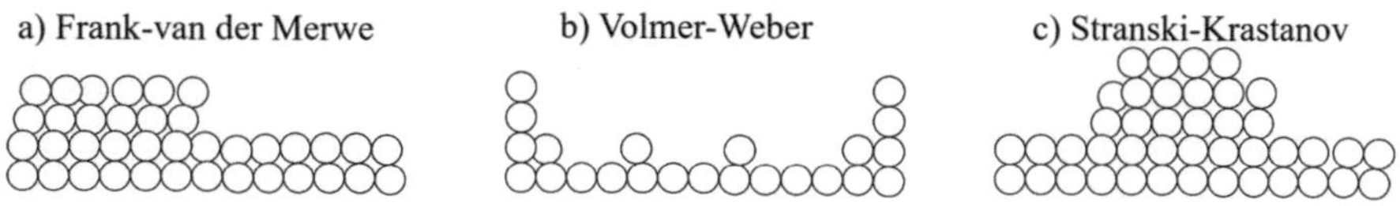

Fig. 7.2 Thin film formation mechanisms

Stranski–Krastanov model depicts a scenario where the interaction forces between surface particles and the substrate surface surpass the forces between the particles themselves. This is the direct opposite of the Van der Merwe growth process. The resulting imbalance between these forces leads to the formation of three-dimensional layered structures on the surface [1, 11].

Various aspects influence thin film deposition processes, which can be classified into three main categories: parameters related to (i) production equipment, (ii) material properties, and (iii) other factors. Key characteristics associated with material attributes include the roughness profile of the substrate, the chemical composition of the target and precursor, and the binding energy of atoms in the target material. The deposition process itself is influenced by several critical parameters, such as the deposition technique, the equipment used, the temperature of the substrate material, the temperatures of the precursor and target materials, the process pressure, the power and type of deposition, the gas flow rate within the deposition chamber, the duration of the process, and the rotational speed of the substrate [1, 12]. Factors beyond those previously mentioned are referred to as external factors. These may include the operator expertise in using the system and the catalysts employed during the process. Achieving high-quality thin films is a complex issue, as multiple parameters are interrelated and must be carefully selected for optimal film formation and deposition. The influence of these parameters on thin film deposition techniques may not always be direct. Therefore, experimental studies and careful design are generally required to determine the optimal parameters for thin film formation [1].

Thin film structures are materials engineered at the nanoscale, which gives them distinct properties compared to bulk materials. These enhanced characteristics can be attributed to their mechanical, optical, electrical, microstructural, and surface properties. The manipulation of atomic organization in thin films can result in improved mechanical properties. As a result, thin films often consist of ceramic compounds that exhibit exceptional hardness, yield strength, modulus of elasticity, and tensile strength. However, traditional mechanical testing methods are not suitable for assessing these materials. Instead, nanoindentation and similar testing equipment are typically used to measure their mechanical properties. Thin film materials can also demonstrate unique reflection and absorption characteristics at various energy levels, offering enhanced optical properties compared to bulk materials. Moreover, in terms of electrical properties, thin films often outperform their bulk counterparts [1, 13].

Thin film materials exhibit a notable scaling effect, where factors such as thickness, lattice parameters, and surface roughness directly influence their electrical properties. The molecular architecture of the synthesized and grown thin film material plays a crucial role in determining its microstructural stability. Some thin film architectures are designed to resist elevated temperatures, while others maintain stability under cryogenic or corrosive conditions. An important characteristic of thin films is their roughness and thickness. The surface roughness of thin films is typically much lower than that of bulk materials and is generally on the nanoscale. The thickness of the thin film is crucial for protecting the substrate surface and ensuring the effectiveness of the desired functional properties. Furthermore, the chemical composition of the thin film can be controlled during the manufacturing process, allowing for the creation of materials with enhanced chemical properties. Surface morphology can be adjusted to improve specific performance characteristics, such as wear resistance, friction, or corrosion resistance, depending on the intended application [6, 7, 14]. One or more processes can generally be used to obtain the desired properties simultaneously; however, not all of them can be achieved at once in every thin film structure. As a result, various characterization techniques are applied to thin film materials. These techniques assess a range of properties, including microstructural, mechanical, chemical, optical, electrical, and magnetic characteristics.

Numerous deposition processes are available for thin film fabrication, with many methods employed to achieve high-quality coatings for engineering applications. To produce high-quality thin films, the deposition process must ensure that the fabricated structures exhibit certain structural qualities. The thin film structures should have smooth interfaces, homogeneous compositions, low defect density, and uniform thickness distribution [1, 6, 13]. When the source material is transformed into a gas state, these procedures are referred to as gaseous deposition techniques. On the other hand, when the source material is transformed into a liquid state, the processes are classified as liquid deposition techniques. Among the most frequently used deposition methods are PVD techniques.

7.3 Physical Vapor Deposition

PVD is commonly employed in vacuum technologies [7, 15, 16]. In this technique, a material surface is coated with oxides, carbides, and ceramics. Solid materials are vaporized as ions, molecules, or atoms, and then transferred to the target material surface under vacuum conditions, where deposition occurs. Initially, diffusion happens on the substrate material surface, followed by combinations and agglomeration. In the final stage, nucleation and deposition take place [17]. Depending on the intended purpose, this results in protective, functional, or decorative coatings [8, 15, 16, 18]. Figure 7.3 shows a schematic example of a PVD system. Thermal energy is required for evaporation from the source material, which can be generated through various sources such as heated wires, laser beams, electron beams, and molecular beams. When vaporized atoms travel significantly in the vacuum and contact the base metal, corrosive liquids can cause more substantial damage. By using an electron beam, atoms are vaporized from the source material, and these vaporized atoms move toward the surface of the material to complete the coating process [19].

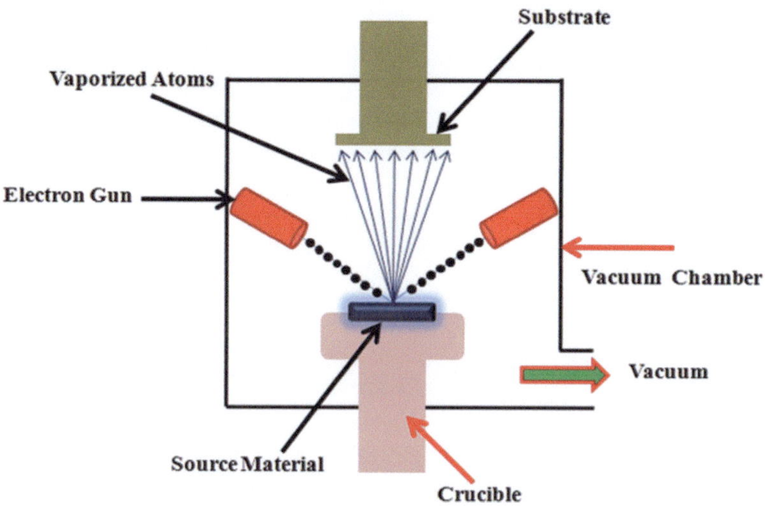

Fig. 7.3 Schematic representation of a PVD system [20]

Figure 7.4 shows the fundamental algorithm used in PVD applications. For successful adhesion of thin films, thorough cleaning and surface preparation of the substrates are essential. Once prepared, the substrate is placed in the vacuum chamber, where preheating is permissible. Due to the elevated temperatures, the target material transitions into the vapor phase, creating a vapor cloud within the evacuated chamber. Vapor-phase atoms are directed toward the substrate using various methods. As they accumulate on the surface, a thin film coating is formed. The vapor clouds can be generated through different methods, but the basic formation process remains the same. The strength and adhesion characteristics of the PVD coating may vary depending on the technology employed [6, 7, 21, 22].

The target surface releases vaporized atoms, ions, or molecules when the binding energy of the target is surpassed, leading to evaporation and ejection from the surface. Thermal heating and sputtering are commonly employed methods for material evaporation. These vaporized components (molecules, ions, or atoms) are then conveyed to the substrate surface through thermal scattering or molecular flow. Collisions occur between the gaseous vapor forms of the plasma and the target substances. Ideally, the target material would achieve 100% efficiency, meaning it would be collisionless. Transferred atoms, ions, or molecules lose velocity upon colliding with the substrate surface, leading to the formation of thin film structures through diffusion, coalescence, nucleation, and condensation. PVD technology allows for rapid coating operations, with higher deposition speeds compared to alternative methods, particularly for metallic surfaces. It is well-suited for coating mass-produced components and can be applied to various target and substrate materials. Furthermore, PVD can be performed at low temperatures, making it applicable to polymer surfaces as well [6, 15, 23].

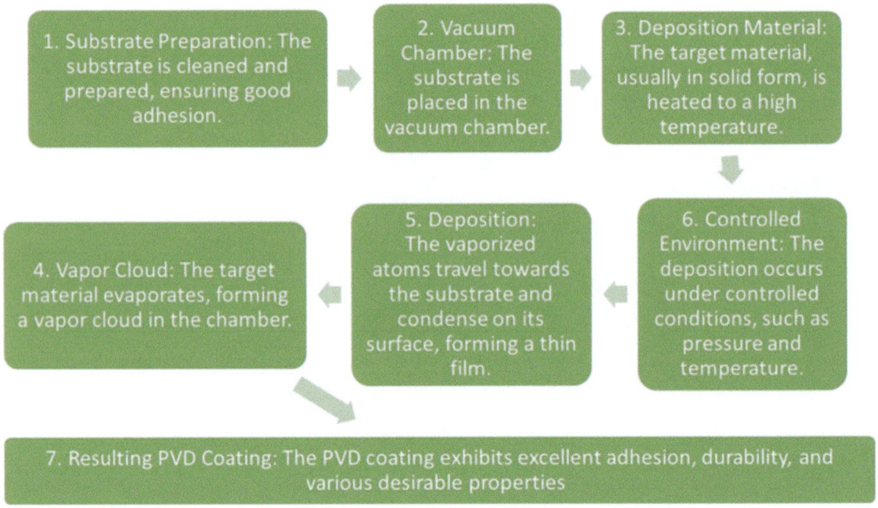

Fig. 7.4 Fundamental algorithm in PVD applications [9]

PVD applies to high-speed steel surfaces that require superior tribological performance, as deposition can occur at low temperatures. It is also suitable for dielectric materials, metals, and nonmetallic substances as targets. Metastable thin film materials, which are advantageous at low temperatures, can be produced using PVD techniques, as the process can be performed under thermodynamic equilibrium conditions [15, 17, 24, 25]. Coatings produced with PVD technology can be precisely regulated in terms of atomic-level coating, film structure, density, and stoichiometry, allowing for exceptional bonding to the substrate material. Coated surfaces exhibit excellent resistance to corrosion and abrasion [17]. Furthermore, because PVD processes occur at low temperatures, they do not induce stress in the substrates, unlike CVD methods [17, 26]. This approach enables the combination of materials with varying compositions, and a wide range of coatings can be manufactured. Multiple target materials may be used to create thin films composed of alloys, high entropy structures, oxides, and nitrides. The impurity levels in the resulting thin film materials are lower than those produced by alternative processes. Due to its previously mentioned attributes, PVD is applicable across various fields. Applications include photovoltaic cells for solar energy, self-cleaning optical lenses, computer chips, and functional and decorative coatings [14, 17, 27, 28]. It is also utilized in environments where biological and tribological properties are critical [29, 30]. Typically, thin film coatings produced by PVD exhibit enhanced mechanical strength, adhesion strength, hardness, and corrosion resistance [26].

A range of PVD systems adheres to comparable fundamental principles for deposition, although they differ in their vapor production techniques. Sputtering and thermal evaporation are the dominant and widely utilized fundamental PVD processes. Numerous variations of these two fundamental techniques have been implemented to

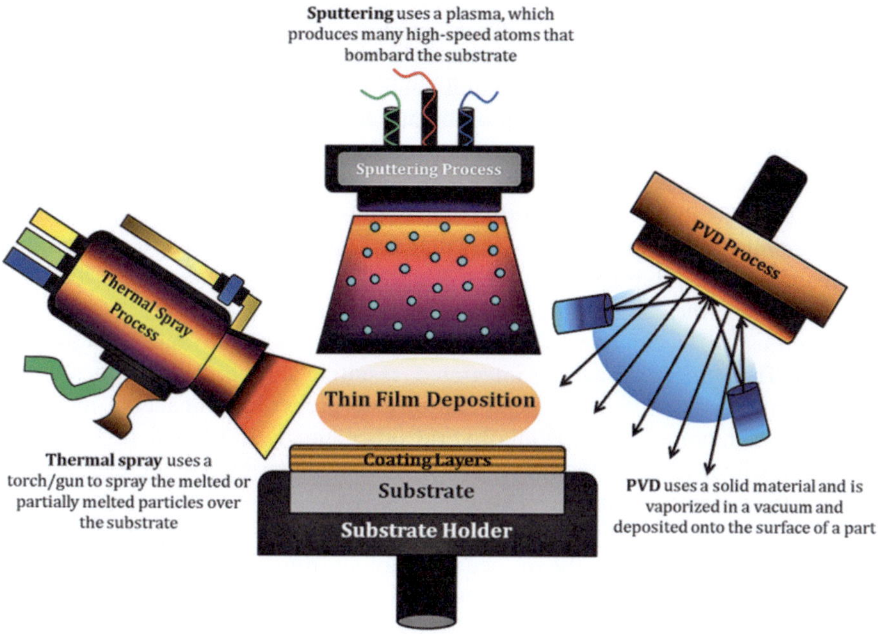

Fig. 7.5 Common thin film deposition methods in applications [9]

enable the creation of diverse coatings on various substrates. Different operating principles exist for these processes [31]. Figure 7.5 shows prevalent thin-film deposition methods for high-temperature, wear-resistant, and corrosion-resistant coatings.

Different techniques are being developed in a multifaceted way to improve the properties of PVD-type thin films. New-generation PVD techniques, modified for the desired properties, have been developed. The currently developed techniques are discussed in the following sections.

The magnetron sputtering process is one of the most frequently employed methods in thin film deposition. This approach utilizes magnetic fields, with the target coating positioned near the applied surface. The magnetic field induces ionization of the sputtered material, which enhances the deposition rate. This technique is known for its substantial deposition rate and is effective in creating complex structures, such as multilayers or gradients [15]. Figure 7.6 depicts a schematic representation of the magnetron sputtering process. Electron beam PVD is another method of PVD deposition. This process involves thermally heating and vaporizing the substance to be coated using an electron beam. The coating parameters are precisely regulated based on the electron beam and processing temperatures. This method is particularly useful for depositing ceramics onto surfaces, especially those requiring elevated temperature input [15]. It offers efficient film characteristics and enhanced functionality. Figure 7.7 shows the schematic operational mechanism of the electron beam PVD process. Another PVD method employed is the pulsed laser PVD deposition technique. This approach involves vaporizing the target substance using a high-powered laser beam. The pulsed laser generates a reactive plasma, facilitating the deposition of diverse materials and complex structures as thin films. The resultant thin films typically exhibit superior microstructural quality compared to other methods [15]. Figure 7.8 is a schematic illustration of the pulsed laser PVD deposition technique. The arc evaporation PVD process generates a high-current electric arc that induces the evaporation of the target substance. This results in the production of a high-density plasma, which facilitates the deposition of thin films. Thin film coatings produced by this technology offer superior adhesive strength. This approach is particularly useful for creating stacked thin film structures. Figure 7.9 shows a schematic of the PVD method utilizing the cathodic arc evaporation technology [15].

Numerous PVD techniques are available, each with distinct advantages and disadvantages. These methods can be integrated, resulting in hybrid techniques aimed at providing coatings with advanced and diverse properties, enhancing industrial processes. A representation of this hybrid procedure is illustrated in Fig. 7.10. The process involves the integration of magnetron sputtering and arc evaporation, enabling the creation of multilayer structures [15].

Various PVD processes produce thin film coating structures with identical chemical compositions. However, the surface morphology, roughness, and cross-sectional microstructure of the thin films created by different processes may vary, significantly

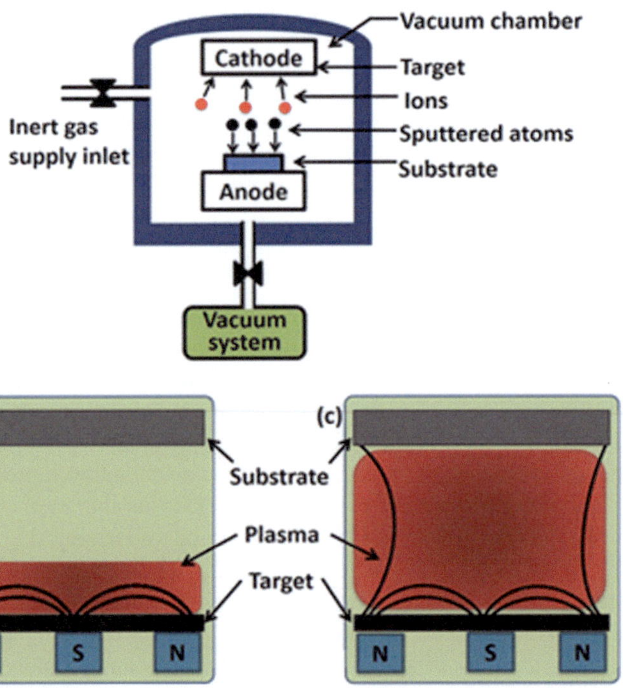

Fig. 7.6 Schematic illustration of PVD magnetron sputtering process [8]

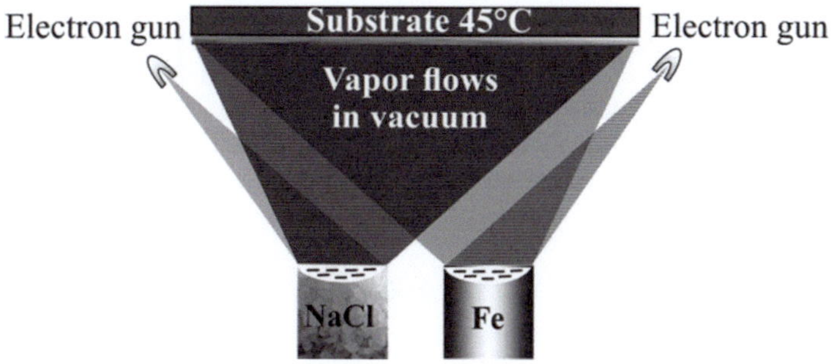

Fig. 7.7 Schematic illustration of electron beam PVD process [32]

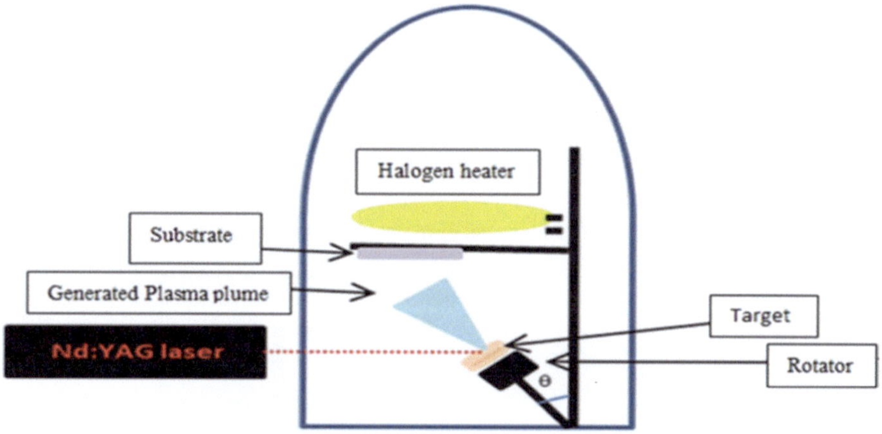

Fig. 7.8 Schematic illustration of pulsed laser PVD deposition process [33]

influencing their mechanical and tribological properties. For example, Yang et al. [36] investigated titanium thin film coatings using direct current sputtering, pulsed sputtering, and cathodic arc evaporation methods. As shown in Fig. 7.11, coatings created via direct current sputtering exhibit columnar structures in their cross-sections, with cauliflower-like surface morphologies and visible cavities. In contrast, thin films produced by pulsed sputtering and cathodic arc evaporation show higher density and superior bonding at the coating interface. Moreover, these two methods outperform direct current sputtering in terms of surface adherence [15, 36, 37]. Although thin film coatings with the same chemical composition can be deposited onto the same substrate, the chosen deposition method can significantly alter several key properties, such as surface bonding behavior, cross-sectional and surface microstructure, and mechanical characteristics. These variations in deposition techniques impact the overall performance and quality of the thin film coatings.

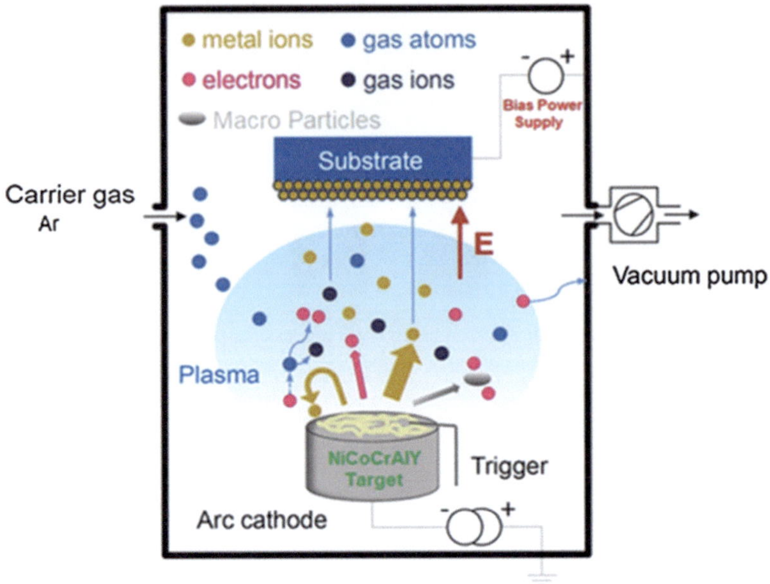

Fig. 7.9 Schematic illustration of cathodic arc evaporated PVD process [34]

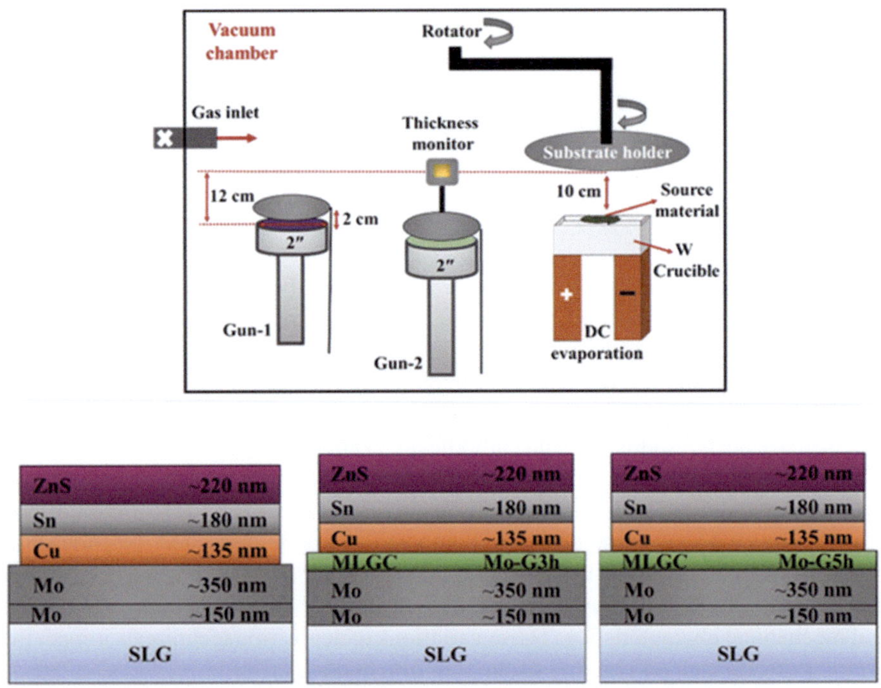

Fig. 7.10 Schematic illustration of a hybrid PVD system and precursor stacks [35]

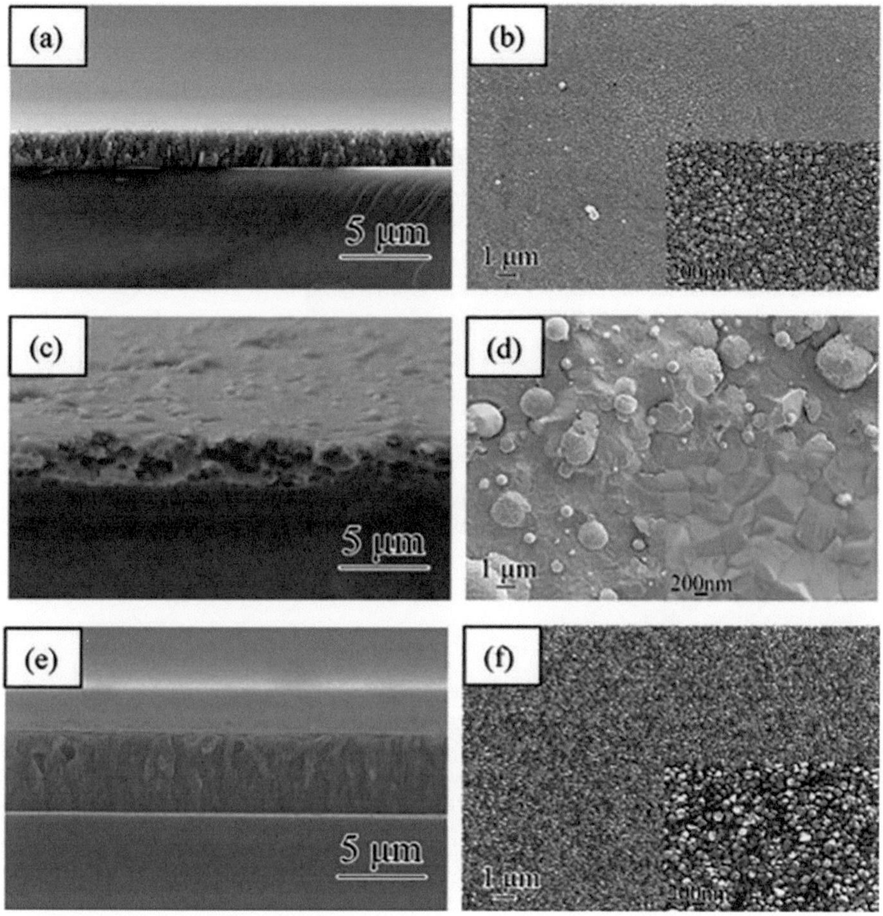

Fig. 7.11 SEM images of titanium thin films deposited in **a**, **b** direct current sputtering, **c**, **d** cathodic arc evaporation, and **e**, **f** pulsed magnetron sputtering processes [36]

7.4 Physical Vapor Deposition for Cutting Tool Coatings

PVD coating processes are widely used across multiple industries, both in everyday tools and complex machine components. These coatings enhance the aesthetic, optical, thermal, and tribological properties of the applied equipment. Various technologies and equipment are utilized within the PVD method, offering superior adhesion strength compared to electroplating techniques. Moreover, PVD coatings provide resistance to oxidation and corrosion. The process allows for the deposition of different nonferrous elements (such as Al, Au, Cu, Ti, Zr), carbide- or oxide-based ceramics (like TiC, TiN), and even materials like stainless steel onto surfaces. PVD is applicable to metallic materials, polymers, glass, ceramics, iron-based, and nonferrous materials. PVD techniques are employed in electronic circuit

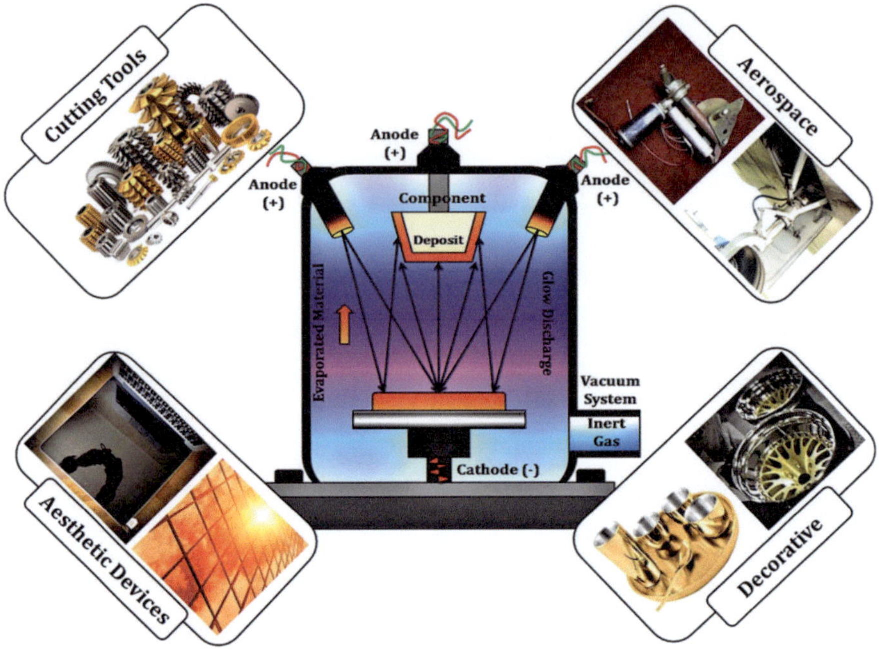

Fig. 7.12 PVD processes in various industrial applications [6]

assemblies, semiconductor components, optical systems, and cutting tools, where aesthetic considerations are important [6, 12, 38]. Figure 7.12 shows the various industrial applications that benefit from PVD processes.

The coating of cutting tools plays a significant role in machining operations by influencing tool longevity, the surface finish of the machined product, and overall production costs. As a result, the cutting tool industry closely monitors and adopts various thin film materials and new technological advancements. In recent years, substantial improvements have been observed in machining efficiency, precision, and the final surface quality of workpieces. These advancements have led to more durable cutting tools, reduced wear rates, and enhanced performance, ultimately optimizing both the quality of the output and cost-effectiveness of the manufacturing processes [38–40].

Machine tools are specifically designed for high-speed machining operations, where they operate at elevated spindle speeds, precise feed rates, and under varying loading conditions [41]. High-speed machining techniques are widely used in industries like aviation, automotive manufacturing, and micromachining, which require high precision for components such as those made from aluminum alloys for the aerospace sector [40]. These advancements in machining technology have led to the development of tools capable of processing not only hard materials but also nanofeatured substances with exceptional precision [40, 42]. In these high-speed machining environments, the cutting tool material must have superior hardness compared to the material being machined. The cutting process generates significant heat due to the friction between the tool and workpiece, which leads to thermal stress. This heat can induce tool wear and compromise the quality of the machined surface finishes, ultimately degrading the tool's performance. To mitigate this issue, coatings are applied to cutting tools to enhance their surface properties, especially their resistance to high temperatures and wear during chip removal [38, 43]. One of the primary benefits of thin film coatings for cutting tools is the enhancement of hardness. Hardness is a critical property for cutting tools, as it ensures the tool can effectively cut through tougher materials without becoming deformed or worn out. Friction during cutting generates heat, which can lower the hardness of the tool material. Coatings made from ceramics such as carbides and nitrides, which are known for their exceptional hardness, are commonly used to combat this problem. These coatings provide superior protection, maintaining the tool hardness even under high-temperature conditions [24, 44–47].

In addition to hardness, the tribological forces acting on cutting tools are crucial to consider. During the cutting process, abrasive contact is a key factor that initiates the frictional forces at the tool–workpiece interface. This frictional behavior, especially under high load conditions, accelerates the wear process. Thin film coatings can provide significant protection to the substrate material by acting as sacrificial layers. They help reduce wear by preventing direct contact between the tool and the workpiece, inhibit material deposits on the cutting edge, and even generate lubricating effects, which mitigate frictional resistance and prolong tool life [46]. Another critical aspect enhanced by thin film coatings is thermal performance. Coatings with high thermal conductivity can efficiently dissipate the heat generated during the cutting process. This is crucial, because excessive heat can lead to thermal softening, reducing the effectiveness and longevity of the cutting tool. Thin films with high thermal conductivity help maintain the tool hardness and performance under high-temperature conditions, ensuring optimal cutting efficiency [48–50]. The stability of the cutting tool surface is also improved with the application of thin film coatings. As temperatures increase during machining, the surface can experience a decline in thermal stability, leading to various detrimental tribochemical reactions, such as oxidation or corrosion. Coatings made from carbides and nitrides offer excellent thermal stability and exhibit ceramic-like properties, which are highly resistant to these reactions. These coatings not only protect the tool surface but also enhance the overall longevity and effectiveness of the cutting tool by preventing material loss due to oxidation or corrosion [46, 48, 51]. Furthermore, thin film coatings significantly enhance mechanical properties, such as fatigue resistance and fracture toughness. These characteristics are critical for tools subjected to repeated mechanical stresses during the cutting process. By improving the ability to withstand cyclic loading and resist cracking, the thin film coatings ensure better overall performance and longer tool life in demanding machining operations.

PVD technique is indeed the leading method for producing thin film coatings with highly specific properties. Among the various coatings developed, nitride-based cutting tools are particularly prominent due to their superior mechanical and tribological properties. Each of these coatings offers distinct advantages depending on the specific requirements of the application. Titanium nitride (TiN) is one of the most widely used coatings, known for its exceptional hardness and wear resistance. This makes it ideal for cutting tools that require high durability and resistance to wear during machining operations. TiN coatings are extensively utilized in the cutting tool industry, because they also provide a smooth, shiny surface, improving performance and reducing friction. Chromium nitride (CrN) coatings, on the other hand, are favored when corrosion resistance is a priority. CrN provides enhanced protection against corrosive environments, making it an excellent choice for tools used in harsh chemical or moisture-laden settings, such as in molds and automotive components. Titanium carbonitride (TiCN) is another variation, offering improved corrosion resistance and mechanical properties. TiCN coatings are particularly beneficial for applications that demand a combination of both strength and corrosion resistance, such as in high-performance machining and cutting tools. Aluminum titanium nitride (AlTiN) coatings are used in high-temperature applications due to their superior thermal stability and wear resistance. AlTiN coatings are particularly effective for machining materials at elevated temperatures, as they can withstand thermal degradation, making them ideal for high-speed and high-heat cutting processes. Diamond-like carbon (DLC) coatings are employed where minimal friction and outstanding wear resistance are required. DLC is known for its low friction coefficient and can be used in applications where reduced wear and tear on both the tool and the workpiece are paramount. These PVD coatings not only improve the performance of cutting tools, but they also extend the lifespan of molds, automotive components, and biomedical applications [9, 52, 53]. The development of new types of coatings continues to be an active area of research, as industries demand materials with increasingly specialized properties for more challenging environments. For example, TiN remains a popular and affordable choice due to its well-balanced performance in terms of hardness, wear resistance, and cost-effectiveness. Meanwhile, CrN is increasingly used in applications requiring both wear and corrosion resistance, highlighting the versatility of PVD coatings across different sectors [8, 38].

The efficacy of binary coatings may be limited in terms of oxidation resistance. For example, TiN undergoes oxidation at temperatures between 450 and 500 °C, which leads to material degradation. Introducing aluminum into the TiN structure at varying rates can improve the oxidation resistance of the thin film coating up to 800 °C. The properties of the resulting ternary nitride structure can be attributed to various compounds. TiN, which has a face-centered cubic structure, enhances wear resistance, while Al_2O_3 in the lower layers provides significant adhesion strength [9, 54]. A similar rationale applies to thin film coatings of CrAlN. The addition of aluminum improves the hardness and wear resistance of CrN coatings, offering superior thermal stability and tribological performance compared to the CrN structure. The constituent elements in CrAlN generate Al_2O_3 and Cr_2O_3 phases at elevated temperatures. This enables the coating to maintain oxidation resistance up to 900 °C, leading to a more robust structure for cutting tool materials [23, 38].

Another example of ternary nitride coatings is the hybrid coatings of TiN and CrN, which incorporate chromium metal into the structure. Compared to binary nitride thin films (TiN and CrN), TiCrN coatings exhibit enhanced hardness, wear resistance, oxidation-corrosion performance, and low friction behavior. Ternary nitride coatings can be fabricated using various PVD processes, such as magnetron sputtering and cathodic arc evaporation, and they typically achieve hardness levels ranging from 30 to 40 GPa [55]. These coatings, a ternary variant of binary CrN thin films, are commonly applied to high-speed steels and cutting tools. Ternary nitride coatings consist of a defined molar ratio of aluminum, chromium, and nitrogen atoms, and they offer significant hardness and toughness. The incorporation of aluminum into CrN coatings further enhances their hardness and wear resistance [56]. These coatings exhibit improved tribological and hardness properties while also demonstrating reduced heat conductivity compared to CrN coatings [23, 38, 57]. Figure 7.13 shows cross-sectional SEM images of different thin film coatings applied to cutting tools.

Another thin film coating commonly used in cutting tool materials is TiSiN, a ternary nitride coating that exhibits exceptionally high hardness levels. Due to its superior characteristics, TiSiN is widely applied in coating cutting tool materials used for machining nonferrous substances [38, 55, 58]. The incorporation of silicon enhances the hardness and wear resistance of TiN, offering superior tribological and oxidation resistance. Ghufran et al. [55] indicated that a silicon content of 7–12% optimizes the mechanical and tribological properties of TiSiN coatings. Quaternary nitride structures, such as TiSiAlN, are another significant thin film material. The addition of aluminum to the structure improves wear resistance while also enhancing ductility and enabling a low-friction regime.

Carbides represent another category of thin film materials used in cutting tools. At their core, carbides are ceramic formations connected by ionic bonds, which contribute to their elevated melting points and significant hardness. These materials

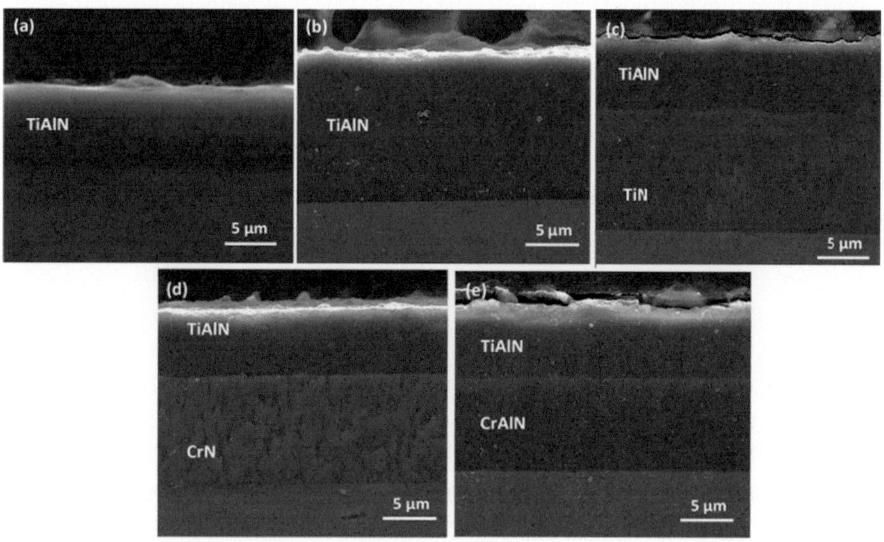

Fig. 7.13 Cross-sectional images of different thin film coatings; **a**, **b** TiAlN, **c** TiN/TiAlN, **d** CrN/TiAlN, and **e** CrAlN/TiAlN [23]

also exhibit excellent electrical and thermal conductivity. TiC and tungsten carbide (WC) are among the most commonly utilized carbides for thin film coatings in cutting tools [59–61]. TiC-based coatings are widely used in steel machining due to their high thermal conductivity, which provides excellent thermal stability. Moreover, TiC coatings offer exceptional wear resistance and chemical stability [59]. However, their corrosion resistance is limited when exposed to acids and halogens, such as HNO_3 and HF. Despite this, TiC coatings exhibit minimal oxidation in air, even at temperatures up to 800 °C, making them particularly suitable for cutting tips used in demanding cutting applications. Another important carbide material is CrC, which is widely used in high-temperature processes such as nitride compound coatings. CrC coatings have hardness values ranging from 3.6 to 48 GPa, depending on their chemical composition [62]. Experimental studies have shown that CrC thin film coatings can enhance wear resistance by up to 40% [63]. Due to their high oxidation and corrosion resistance, CrC coatings are also suitable for applications such as plastic injection molding [64].

A significant advancement enabled by thin film techniques is the development of layered structures. Thin films can exhibit a variety of microstructural characteristics, including monolayer, bilayer, multilayer, gradient, nanoscale layers, or nanocomposite structures [65, 66], as shown in Fig. 7.14. Monolayer structures consist of a single composition, while bilayer structures feature two distinct coatings separated by a boundary. Gradient structures are defined by variations in chemical composition along the coat thickness. Multilayer structures involve alternating layers of different materials, such as carbides and nitrides. Nanoscale layered coatings are termed nanolayered structures, where the layers are on the order of nanometers in thickness. Nanocomposite materials represent the most promising direction for innovation in thin film coating technology. The development of nanocomposite thin film coatings is advancing rapidly, enabling the incorporation of various borides, carbides, nitrides, and metals within the coating [65, 66]. Examples include TiC/TiB_2, TiN/TiB_2, B_4C/W, and SiC/Al. This design strategy fundamentally integrates rigid and flexible materials, utilizing a combination of these materials to optimize the thin film properties. As a result, many thin film coatings designed for enhanced machining performance are fabricated in nanocomposite or nanolayered configurations [38].

7.5 Conclusions

Developing thin film coatings using PVD technique, a microdeposition technology, has become a significant area of focus due to their superior performance characteristics. Thin film coatings are widely utilized on machine components for enhanced functionality, particularly in cutting tools. This chapter presents the fundamentals of thin film deposition techniques and discusses thin film coatings specifically applied to cutting tool surfaces, focusing on their primary application domains. PVD techniques enable the fabrication of various ceramic structures, including nitrides, carbides, and elemental materials, which demonstrate enhanced microstructural, mechanical,

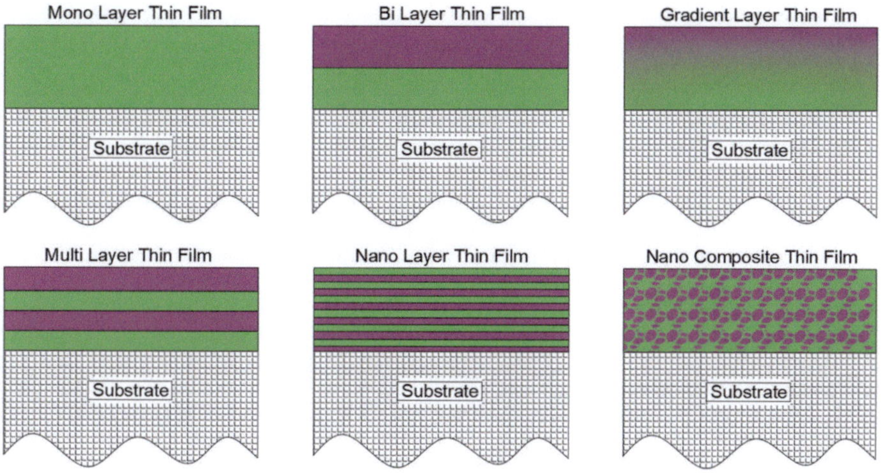

Fig. 7.14 Schematic illustration of different types of thin film coatings

corrosion-resistant, and oxidative properties. The discussion highlights specialized thin film coatings utilized in cutting tool applications, providing their influence on performance and longevity. Moreover, cross-sectional views of thin film coatings are presented for understanding the microstructural evolution within the thin film coatings on the substrates.

References

1. Mwema, F.M., T.-C. Jen, and L. Zhu, Introduction to Thin Films and Coatings, in Thin Film Coatings. 2022. p. 1–15.
2. Wallin, E., Alumina Thin Films: From Computer Calculations to Cutting Tools, in Linköping Studies in Science and Technology. Dissertations. 2008, Linköping University Electronic Press: Linköping. p. 59.
3. Greene, J.E., Review Article: Tracing the recorded history of thin-film sputter deposition: From the 1800s to 2017. Journal of Vacuum Science & Technology A: Vacuum, Surfaces, and Films, 2017. **35**(5).
4. Jilani, A., M.S. Abdel-wahab, and A.H. Hammad, Advance Deposition Techniques for Thin Film and Coating, in Modern Technologies for Creating the Thin-film Systems and Coatings. 2017.
5. Kasap, S., C. Koughia, and P. Capper, Springer Handbook of Electronic and Photonic Materials. 2007: Springer US.
6. Sathish, M., N. Radhika, and B. Saleh, Current Status, Challenges, and Future Prospects of Thin Film Coating Techniques and Coating Structures. Journal of Bio- and Tribo-Corrosion, 2023. **9**(2).
7. Oluwatosin Abegunde, O., et al., Overview of thin film deposition techniques. AIMS Materials Science, 2019. **6**(2): p. 174–199.

8. Atta, S., et al., Recent Developments and Applications of TiN-Based Films Synthesized by Magnetron Sputtering. Journal of Materials Engineering and Performance, 2023. **32**(22): p. 9979–10015.
9. Ichou, H., et al., Exploring the Advancements in Physical Vapor Deposition Coating: A Review. Journal of Bio- and Tribo-Corrosion, 2023. **10**(1).
10. Unold, T. and H.W. Schock, Nonconventional (Non-Silicon-Based) Photovoltaic Materials. Annual Review of Materials Research, 2011. **41**(1): p. 297–321.
11. Levi, A.C. and M. Kotrla, Theory and simulation of crystal growth. Journal of Physics: Condensed Matter, 1997. **9**(2): p. 299–344.
12. Mwema, F.M., et al., Properties of physically deposited thin aluminium film coatings: A review. Journal of Alloys and Compounds, 2018. **747**: p. 306–323.
13. Rao, M.C. and M.S. Shekhawat, A Brief Survey on Basic Properties of Thin Films for Device Application. International Journal of Modern Physics: Conference Series, 2013. **22**: p. 576–582.
14. Mwema, F.M., T.-C. Jen, and L. Zhu, Characterisation Techniques of Thin Films, in Thin Film Coatings. 2022. p. 77–114.
15. Mwema, F.M., T.-C. Jen, and L. Zhu, Methods of Thin Film Deposition, in Thin Film Coatings. 2022. p. 17–76.
16. Martín-Palma, R.J. and A. Lakhtakia, Vapor-Deposition Techniques, in Engineered Biomimicry. 2013. p. 383–398.
17. Deng, Y., et al., Physical vapor deposition technology for coated cutting tools: A review. Ceramics International, 2020. **46**(11): p. 18373–18390.
18. Baptista, A., et al., Sputtering Physical Vapour Deposition (PVD) Coatings: A Critical Review on Process Improvement and Market Trend Demands. Coatings, 2018. **8**(11).
19. Mwema, F.M., E.T. Akinlabi, and O.P. Oladijo, Fractal analysis of hillocks: A case of RF sputtered aluminum thin films. Applied Surface Science, 2019. **489**: p. 614–623.
20. Kumar, N. and V.K. Choubey, Recent trends in coating processes on various AISI steel substrates: A review. Journal of Materials Science, 2023. **59**(2): p. 395–422.
21. Mattox, D.M., Ion plating — past, present and future. Surface and Coatings Technology, 2000. **133–134**: p. 517–521.
22. Friedrich, K., Polymers for Low-Temperature Tribology Applications, in Encyclopedia of Tribology. 2013. p. 2654–2661.
23. Adesina, A.Y., Tribological Behavior of TiN/TiAlN, CrN/TiAlN, and CrAlN/TiAlN Coatings at Elevated Temperature. Journal of Materials Engineering and Performance, 2022. **31**(8): p. 6404–6419.
24. Moghaddam, P.V., et al., High temperature tribology of TiAlN PVD coating sliding against 316L stainless steel and carbide-free bainitic steel. Tribology International, 2021. **159**.
25. Sousa, V.F.C., et al., Study of the wear behaviour of TiAlSiN and TiAlN PVD coated tools on milling operations of pre-hardened tool steel. Wear, 2021. **476**.
26. Koseki, S., et al., Comparison of TiN-coated tools using CVD and PVD processes during continuous cutting of Ni-based superalloys. Surface and Coatings Technology, 2015. **283**: p. 353–363.
27. Subedi, D.P. and G.P. Panta, Electrical characterization of aluminum (Al) thin films measured by using four- point probe method. Kathmandu University Journal of Science, Engineering and Technology, 2013. **8**(2): p. 31–36.
28. White, J., et al., Environmentally benign metallization of material extrusion technology 3D printed acrylonitrile butadiene styrene parts using physical vapor deposition. Additive Manufacturing, 2018. **22**: p. 279–285.
29. Ehiasarian, A.P., et al., Comparison of microstructure and mechanical properties of chromium nitride-based coatings deposited by high power impulse magnetron sputtering and by the combined steered cathodic arc/unbalanced magnetron technique. Thin Solid Films, 2004. **457**(2): p. 270–277.
30. Marchiori, G., et al., Optimizing thickness of ceramic coatings on plastic components for orthopedic applications: A finite element analysis. Mater Sci Eng C Mater Biol Appl, 2016. **58**: p. 381–8.

31. Dan, A., et al., A review on physical vapor deposition-based metallic coatings on steel as an alternative to conventional galvanized coatings. Journal of Coatings Technology and Research, 2022. **19**(2): p. 403–438.
32. Kurapov, Y.A., et al., Electron-Beam Physical Vapor Deposition of Iron Nanoparticles and their Thermal Stability in the Fe–O System. Powder Metallurgy and Metal Ceramics, 2022. **60**(7–8): p. 451–463.
33. Salim, E.T., et al., Polycrystalline T- and H-Nb2O5 Thin Films Prepared by Pulsed Laser Deposition: Impact of Laser Fluence. Journal of Electronic Materials, 2024. **53**(10): p. 6482–6497.
34. Kheyrodin, M., F. Mahboubi, and H. Elmkhah, The Effect of Negative Bias Voltage on the Microstructure and Hot Corrosion Behavior of Heat-treated NiCoCrAlY Coatings Applied Via the Cathodic Arc Evaporation Physical Vapor Deposition Method. Journal of Materials Engineering and Performance, 2023. **33**(1): p. 283–300.
35. Akcay, N., et al., Cu2ZnSnS4 films prepared by a hybrid PVD deposition system: a multi-layered graphitic carbon intermediate layer at the Mo/CZTS interface. Journal of Materials Science: Materials in Electronics, 2024. **35**(32).
36. Yang, C., et al., Structure and properties of Ti films deposited by dc magnetron sputtering, pulsed dc magnetron sputtering and cathodic arc evaporation. Surface and Coatings Technology, 2016. **304**: p. 51–56.
37. Hom-on, C., et al., Surface roughness of aluminum oxide thin films deposited by DC and RF reactive magnetron sputtering. Materials Today: Proceedings, 2018. **5**(7): p. 15228–15232.
38. Mwema, F.M., T.-C. Jen, and L. Zhu, Thin Films for Cutting Tools, in Thin Film Coatings. 2022. p. 163–182.
39. Hashmi, K.H., et al., Optimization of process parameters for high speed machining of Ti-6Al-4V using response surface methodology. The International Journal of Advanced Manufacturing Technology, 2015. **85**(5–8): p. 1847–1856.
40. Jain, A. and V. Bajpai, Introduction to high-speed machining (HSM), in High Speed Machining. 2020. p. 1–25.
41. Dewes, R.C., D.K. Aspinwall, and M.L.H. Wise, High Speed Machining — Cutting Tools and Machine Requirements, in Proceedings of the Thirty-First International Matador Conference, A.K. Kochhar, Editor. 1995, Macmillan Education UK: London. p. 455–461.
42. Hon, K. and B.T.H.T. Baharudin, The impact of high speed machining on computing and automation. International Journal of Automation and Computing, 2006. **3**(1): p. 63–68.
43. Narasimha, M., Improving Wear Resistance of Cutting Tool by coating. IOSR Journal of Engineering, 2014. **4**(5): p. 06–14.
44. Piska, M., Hard Nano-Crystalline Coatings for Cutting Tools. Materials Science Forum, 2007. **567–568**: p. 185–188.
45. Chowdhury, M.S.I., et al., Investigation of the Wear Behavior of PVD Coated Carbide Tools during Ti6Al4V Machining with Intensive Built Up Edge Formation. Coatings, 2021. **11**(3).
46. Kathrein, M., et al., Doped CVD Al2O3 coatings for high performance cutting tools. Surface and Coatings Technology, 2003. **163–164**: p. 181–188.
47. Henderer, W. and F. Xu, Hybrid TiSiN, CrC/C PVD coatings applied to cutting tools. Surface and Coatings Technology, 2013. **215**: p. 381–385.
48. Shri, D.N.A., et al., Influence of Surface Pretreatment on Carbon Coating of Cutting Tools Using PVD. Applied Mechanics and Materials, 2012. **236–237**: p. 530–535.
49. Naumov Alexander, G., et al., Environmentally Friendly Technological System of Cutting Using Magnetic Microcapsules and Cutting Tools with Nanoscale Composite Coating. Procedia CIRP, 2016. **41**: p. 829–834.
50. Kowalczyk, J., et al., The Influence of Cutting Fluid and Diamond-Like Carbon Coating on Cutting Tool Wear. DEStech Transactions on Computer Science and Engineering, 2019(fe).
51. Posti, E. and I. Nieminen, Influence of coating thickness on the life of TiN-coated high speed steel cutting tools. Wear, 1989. **129**(2): p. 273–283.
52. He, Q., et al., An integrative approach to coating/carbide substrate design of CVD and PVD coated cutting tools during the machining of austenitic stainless steel. Ceramics International, 2020. **46**(4): p. 5149–5158.

53. Guo, X., et al., Controllable preparation of micro-textures on WC/Co substrate surface by an integrated laser-dry etching process for improving PVD coatings adhesion. Applied Surface Science, 2020. **534**.
54. Park, J.-H., et al., Effect of Zn-Mg interlayer on the corrosion resistance of multilayer Zn-based coating fabricated by physical vapor deposition process. Corrosion Science, 2022. **202**.
55. Ghufran, M., et al., Development and tribo-mechanical properties of functional ternary nitride coatings: Applications-based comprehensive review. Proceedings of the Institution of Mechanical Engineers, Part J: Journal of Engineering Tribology, 2020. **235**(1): p. 196–232.
56. Yu, C.Y., et al., Tribological behaviour of CrAlN coatings at 600°C. Surface Engineering, 2013. **29**(4): p. 318–321.
57. Li, P., et al., Microstructure, mechanical and thermal properties of TiAlN/CrAlN multilayer coatings. International Journal of Refractory Metals and Hard Materials, 2013. **40**: p. 51–57.
58. Wang, C.Y., et al., Wear and breakage of TiAlN- and TiSiN-coated carbide tools during high-speed milling of hardened steel. Wear, 2015. **336–337**: p. 29–42.
59. Santerre, F., et al., Properties of TiC thin films grown by pulsed laser deposition. Applied Surface Science, 1999. **148**(1–2): p. 24–33.
60. Badaluddin, N.A., et al. Coatings of Cutting Tools and Their Contribution to Improve Mechanical Properties: A Brief Review. 2018.
61. Phiri, R.R., O.P. Oladijo, and E.T. Akinlabi, Tungsten carbide thin films Review: Effect of deposition parameters on film microstructure and properties. Procedia Manufacturing, 2019. **35**: p. 522–528.
62. Li, Z.-L., et al., Comparison of chromium carbide thin films grown by different power supply systems. Surface and Coatings Technology, 2018. **353**: p. 329–338.
63. Singh, V., et al., Chromium carbide–CNT nanocomposites with enhanced mechanical properties. Acta Materialia, 2009. **57**(2): p. 335–344.
64. Su, Y.L., et al., Wear of CrC-coated carbide tools in dry machining. Journal of Materials Processing Technology, 2006. **171**(1): p. 108–117.
65. Sousa, V.F.C. and F.J.G. Silva, Recent Advances on Coated Milling Tool Technology—A Comprehensive Review. Coatings, 2020. **10**(3).
66. Sidorenko, D., et al., Nanocomposites for Machining Tools. Materials (Basel), 2017. **10**(10).

MIX
Papier aus verantwortungsvollen Quellen
Paper from responsible sources
FSC® C105338

If you have any concerns about our products,
you can contact us on
ProductSafety@springernature.com

In case Publisher is established outside the EU,
the EU authorized representative is:
Springer Nature Customer Service Center GmbH
Europaplatz 3, 69115 Heidelberg, Germany

Printed by Libri Plureos GmbH
in Hamburg, Germany